40 Weeks

40 Weeks

What Humans and
81 Other Species Expect
When They're Expecting

ANNA BLIX

Translated from the Norwegian by
Nichola Smalley

MACLEHOSE PRESS
QUERCUS · LONDON

First published in Norwegian as
40 uker: En menneskegraviditet og 81 andre måter å få barn på
by Cappelen Damm As, Oslo, in 2023
First published in Great Britain in 2025 by

MacLehose Press
An imprint of
Quercus Editions Limited
Carmelite House
50 Victoria Embankment
London EC4Y 0DZ

An Hachette UK company

The authorised representative in the EEA is Hachette Ireland,
8 Castlecourt Centre, Dublin 15, D15 XTP3, Ireland (email: info@hbgi.ie)

Copyright © Anna Blix, 2023
Translation copyright © 2025 by Nichola Smalley
Illustrations by Frøydis Sollid Simonsen

This translation has been published with the financial support of NORLA

The moral right of Anna Blix to be
identified as the author of this work has been
asserted in accordance with the Copyright,
Designs and Patents Act, 1988.

Nichola Smalley asserts her moral right to be identified as the translator of this work.

All rights reserved. No part of this publication
may be reproduced or transmitted in any form
or by any means, electronic or mechanical,
including photocopy, recording, or any
information storage and retrieval system,
without permission in writing from the publisher.

A CIP catalogue record for this book is available
from the British Library

ISBN (HB) 97 81 52943 480 4
ISBN (TPB) 978 1 52943 481 1
ISBN (EBOOK) 978 1 52943 483 5

Quercus Editions Ltd hereby exclude all liability to the extent permitted by law for any errors
or omissions in this book and for any loss, damage or expense (whether direct or indirect)
suffered by a third party relying on any information contained in this book.

1

Typeset in Albertina MT by CC Book Production
Printed and bound in Great Britain by Clays Ltd, Elcograf S.p.A.

Papers used MacLehose Press are from well-managed forests and other responsible sources.

40 Weeks

Week 41

"If you reach down now, you'll feel his little head. He's got masses of hair!"

All my muscles strain. My body has been working towards this moment for so long. Now, just as the sun is rising above the horizon again, comes the culmination of the last few hours' effort and the last few months' gradual construction of a little body within my own. Until last night I was a rational person with thoughts and opinions about how the birth would be. Now I'm an animal, acting on instinct.

The person in me disappeared as the contractions grew in strength. They have gone from being a tickle in the small of my back to waves that wash right through me. I snarled at the midwife until she left me in the bathroom, doubled over the back of a chair in the shower, the hot water washing away my pain. I've been gracious enough to let my partner give me the occasional sip of water, but I'm interested in neither talking nor hugs. My body is giving birth to a baby, and my brain's just along for the ride.

*

I realised a few hours ago that tonight was the night. I was lying on the sofa watching TV when the practice contractions I had been having (these are pains that can happen throughout late pregnancy and help train the uterus) suddenly started to grow more intense. There was a cushion between my knees and a cup of tea on the coffee table, and the book I'd been intending to read lay discarded on the floor – yet another evening when I wouldn't be able to concentrate on the words. Yet another evening like so many these past weeks. An evening I'd expected to involve little more than me trying not to move my pelvis too much, resting so I'd be able to walk the following day, feeling the movements of the foetus, and wondering when it would finally happen.

The contractions got stronger and closer together, and it became more difficult to talk through them. We booked a taxi, thinking, since it was a Friday night, that it would be smart to get to the hospital before people started heading home from town and it got harder to find a cab. In the taxi it was like my brain had given up controlling my body, as though it no longer had the wherewithal to make me behave within the bounds of social norms. My groans came out through clenched teeth, and I had to shout my way through the next contraction. I was clinging to the hand-hold above the window on the back seat as the driver went faster and faster, probably scared I was going to give birth in his cab.

We got to the hospital, and on the forecourt I had a contraction so strong I had to bend double before I was able to walk in. I was hanging off my partner, who supported me with one arm while he carried the bag with all our things in his other hand. In the lift I breathed my way up the levels. All I wanted was to get into the delivery room, to see if it was like the one I gave birth in last time, to familiarise myself with my surroundings, smell the bed linen and make sure everything was safe. My brain came back to life a little when I met the midwife; it was Torild, who I'd met before. I'd known she'd be on duty this evening, and that comforted me: I trusted her. My brain retreated again as I undressed. I didn't have it in me to answer Torild's question about how I saw the evening unfolding. All I wanted was to get in the shower, to feel the water running down my back, to dull the pain with warm streams of ancient sea.

When the pains got so intense I was about to give up, when the warm water came to feel like a sticking plaster on an open fracture, I waddled over to the bed so the midwife could check how the birth was progressing. "Turn off the light," was all I managed to say. Like an animal I needed to give birth in darkness, in safety; concealed from the predators out on the savanna. She stuck her fingers inside me, and my brain had to remind my body that she was a friend, a helper, not an attack on the birth or my cervix. "Eight centimetres," I heard her say as I growled. She withdrew, giving me the security of her closeness while she let my body do its work. I could feel the sweat running down my chest, the rhythmic bracing of my muscles, my uterus pushing the baby further and further down with each contraction. I felt his head in the birth canal, how with each push he slipped first downwards and then back up a little. Two steps forward and one step back, a bit further each time. My body can handle this on its own; my mind is confident: it's time for my uterus to do its job.

"Now, pant like a dog! Don't push!" His head is out and he gives a little whimper, the only noise he's capable of making while his lungs are compressed during his body's transition from umbilical cord to air. We laugh a little, all of us, at this strange sound from a baby's head with no body, halfway out of me. I see myself from the outside – the animal has retreated and now I can think and talk again. I'm almost done, my body knows the worst is over. I touch the damp hair. "Look, you'll soon be able to hold him yourself. He'll be out on the next contraction." The midwife is speaking in a calm voice. My partner's holding my hand. My body strains, I reach down. Then he's there, on my stomach. He draws air into his lungs for the first time, and this hard, cold air he's never encountered before hits his body simultaneously from the inside and the outside. His skin dries for the first time in his life as we wipe away the amniotic fluid that has encompassed and comforted him until now. The umbilical cord is still pumping blood into him, like a lifeline from the safety of the uterus. It will soon be cut, and our nine-month symbiosis will be replaced by a new phase.

He lies there, three and a half minutes – and 3.5 billion years – old. He's the primordial soup, and he's a completely new life that has never existed before. He can do nothing – the most helpless infant of any species – except recognise my voice, smell my milk and creep his way from my belly to my nipple.

He lifts his enormous head, much too heavy for his little body, and stares at my face to imprint my features on his memory. He opens his mouth and encircles my nipple with the world's strongest vacuum. He's been inside me for nine months and I am finally released from my role as his place of residence, but that sucking latch makes it clear I'm not done yet – my body is still subject to his needs.

I'm no longer the dwelling place of a little organism that's given me side effects on a level with a serious parasite, like an alien lifeform sponging off my body.[1] I have spent the past months vomiting, so sick I was given four different prescription anti-nausea drugs. My inner organs have been re-organised, my gut and bladder compressed to make space. My skin has been stretched as far from my spinal column as it would go, the pregnancy hormones have given me constipation and a need for hours of extra sleep. It's been months since I could tie my own shoelaces, I've been forced to wear support stockings, and I've been munching antacids like nobody's business to counteract the constant heartburn. In order to get my progeny out, my skeleton has begun to come apart, and with every step I've taken, the pain in my pelvis has got worse and worse. As this little organism has grown, my body has borne the brunt of my species' mode of reproduction.

Imagine if I'd been able to simply lay an egg and get my partner to sit on it until it hatched? Or give birth to a baby so little I barely noticed the birth, keeping it in a pouch until it was big enough to face life? That said, at least I don't have to do what the spotted hyena does: give birth through an elongated clitoris shaped like a penis that's so narrow that 60 per cent of the pups of first-time mothers die. I'm not pregnant inside my skeleton, like I would be if I were a scorpion — forced to crawl around like an inflated balloon towards the end, its exoskeleton distended by the young inside its body. I don't have to sit still on a nest until my young hatch, like the common eider. I don't starve to death while guarding my eggs, like the octopus, and I don't let my young eat me alive, like some spiders do.

Right now, with my newborn baby lying on me, slowly creeping his way towards my nipple to take his first gulp of milk, it feels right. My body is high on all the hormones that set labour in motion and provided me with pain relief and are now making sure I fall in love with this shrivelled creature that's come out of me. I'm high on the emotion of having managed to give birth to this cute little baby, and those nine long months seem worth it. But along the way there have been many times when I've wished I belonged to a different species, a different species with a different reproduction solution, any solution but the one that allowed a fertilised egg to embed itself in my tissue, and an embryo to control my blood and take over my body.

All the organisms that came into this world at the very same time as my baby – all the parents who divided themselves, who released unfertilised eggs into the water so they might find the sperm themselves, who saw a little head sticking up out of an eggshell, felt their young squirming in the skin of their back or their vocal sac, or who forced their young out through a narrow birth canal – we're all the furthest out on the tree of life, each on our own branch, but we all come from the same trunk, from that same primordial soup, those first living cells. And we've all survived long enough to produce offspring, whether we're humans, amoebas, sea anemones, hyenas, eiders or kangaroos. On the way to this point, the organisms that came before us changed little by little, each in their own way, and at last here we all are, with an abundance of different methods for reproducing ourselves. This is the story of a few of all the fantastic ways we do it.

Week 1

I got a notification telling me it was about to happen. An app sent me a friendly reminder that a 40–50-million-year-old trait[2] is about to be acted out once again inside my body. But I swipe it away, I can't be bothered to think about my period, and so it is that I wake to blood on my sheets. I don't want blood – it's a clear signal it hasn't happened this month either. I want to have a baby, and that calls for bodily fluids of a different kind. It means mucus, sperm, thick uterine linings, hormonal changes and a swelling tummy. Planning, fucking, waiting.

It's much easier for some. The clonal plumose anemone (*Metridium senile*) looks like a plant but is more closely related to us. It has taken up residence on the post of a jetty – shallow enough that you can see it if you lie with your head poking over the edge, looking down into the water on a warm summer's day.[3] It consists of a long, round body with a large number of tentacles on top, and looks rather like a white anemone flower with a thick, rust-coloured stalk. The plumose anemone – an animal that belongs to the class Anthozoa and is a common sight along the whole coast of Norway – can quite simply grow itself a little genetically identical offspring, like a small lump on the side of its own body. And why not, since it's just standing there sifting the water through its tentacles for food? A bud grows out from the body, a baby sea anemone, which, when it's big enough, with a body and tentacles and a tiny mouth in the middle (the plumose anemone doesn't consist of much more), breaks loose and begins its own life.[4]

Does the anemone notice that a little one is growing out of its body? Has it actively decided that the time is right for it to reproduce? Does it get tired? Does its stalk ache? Keeping one's offspring inside one's body, in that specialised organ, the uterus, is not the only way of doing it. Parents grow and keep their young in their mouths, between their legs, in a large cavity inside their bodies that is also used to digest food, in little cavities in their backs, in fertilised eggs in a guarded nest, directly on their bodies like the sea anemone, or in some cases they simply release their eggs and sperm and let it happen. My method, with a uterus, placenta, and the intimate exchange of resources over many months, is, as we will see, a fairly new invention,[5] but life on earth has been reproducing itself since it first developed, more than 3.5 billion years ago.

There's such infinite variation in how different species pass on their genes, producing generation after generation. Before I've even properly begun my journey, the single-cell bacteria *Escherichia coli* is already done. Some variants of *E. coli* can produce poisons that make us very sick if they get inside us, but it is, in fact, normally present in our guts. When the *E. coli* bacterium has a good habitat with good access to nutrients, as in the human gut, it gets ready to copy itself. The process is called binary fission, and it begins with the bacterium copying its DNA. Once there are two full copies of its genes, the two lengths of DNA move to opposite ends of the cell. The rod-shaped bacterium grows longer and longer, until suddenly, the cell wall joins in the middle. It splits into two halves, which move apart, and in the space of twenty minutes, one bacterial cell has become two. I will take forty weeks. Making a human is a little more complex than making an *E. coli* bacterium. But by the time I've given birth, that one *E. coli* cell could theoretically have become the progenitor of more bacteria than there are atoms in the universe, given sufficient habitats and enough food.[6] You could well ask who has the more successful strategy.

*

But today I'm not splitting in two, I'm bleeding on my sheets. For the last three weeks, my uterus has been enthusiastically laying down a lining of mucous membrane, and the blood stain is the product of that. A mature egg has been flung out of my ovary, along my uterine tube towards my uterus. It tried to embed itself there, but either no sperm cell came along to fuse with it and give it the chromosomal structure that would allow it to do so, or else the fertilised egg and my uterus didn't cooperate. When I bleed, it's because the egg hasn't attached to the endometrium built up by the uterine wall, and the uterus throws away its last few weeks' work. The egg and the endometrium are flushed out; the baby and the placenta – that 140-million-year-old organ, relatively new in evolutionary terms, that would have facilitated the exchange of nutrients between the baby and me – are not created. The uterus, approximately eight centimetres long in my case, then prepares itself to once again receive an egg.

My blood runs down my legs in the shower. Well, at least I won't have to drag around a baby half-grown out of my hip like the sea anemone, even if I would have liked my belly to have a little bulge. The blood creates intricate patterns on the tiles at my feet before it's washed down the drain. My unfertilised egg and the endometrium that would have formed a protective layer around the growing embryo flow out into the sewers, into the sea. Perhaps some sea creature will be able to make use of the nutrients in the egg and blood. My body doesn't want them.

The coral *Lophelia pertusa* lives deep, deep down in the waters of one of the world's largest cold-water coral reefs, near Røst, off the Lofoten Islands of northern Norway. When it procreates, there is no uterus involved. It releases its eggs straight into the sea, shooting out the eggs like a volcano spewing lava, and if we were down on the reef right now, we would have seen them spreading through the water. The males release sperm, which mix with the eggs.[7] How are they able to coordinate the simultaneous release of eggs and sperm? Is the female coral anxious to know whether her eggs get fertilised? Is she sore and bloated as they pass out of her body?

I wipe myself clean and look for a sanitary towel. There are only a few of us, of all the species on earth, that menstruate. But isn't it an enormous waste of resources to bleed out this layer of mucous membrane every month? I dig out my iron tablets. Important elements of the blood running out must be replenished.

As I bleed, the large, colourful cichlid fish *Ctenochromis horei* prepares to lay its eggs in a nest on the bed of the lake it lives in. It has found a sandy spot between the rocks and patches of mud on the bottom of Lake Tanganyika, the large body of fresh water that separates Tanzania and the Democratic Republic of the Congo and is probably the deepest lake in the world. The male *Ctenochromis horei* is multicoloured. He has yellow and black markings on his head, red spots on a silver-grey background along his sides, often interspersed with a little pale blue. The female, who is shorter than the male's fifteen to twenty centimetres, is not as colourful, but still interesting enough to look at that experimental aquarists often keep them. Here, in the lake, in freedom, the male has painstakingly moved sand and stones until he's happy with his nest, a hollow in the sandy lake bed surrounded by green plants that sway in the calm, blue-green waters. It won't be used for long, only for the mating dance: almost as soon as the eggs are deposited, they'll be rerouted to their hatching place. The male starts dancing in front of the female, and at once his colours become brighter.[8] She dances with him and lays a few eggs at a time in the nest. As she does so, he sprays them with sperm, but she doesn't leave it up to his aim to determine whether or not they get fertilised. She immediately picks up the eggs and the sperm using her mouth, where they will be better protected from the water currents and can mix more easily. Her plan is to hold these eggs in her mouth for the next week, until they hatch. There are few fish species with internal fertilisation and incubation, and this species is trying to compensate by using the mouth as a uterus. They are mouthbrooders, using their oral cavities to ensure the eggs don't get eaten by others, something that could easily happen in the densely populated lake they inhabit.

And indeed, there's something lurking nearby that would dearly love to eat the eggs, but that's not the only thing it's after when it interrupts the mating and fertilisation dance of the cichlids. While the cichlid couple are cavorting, a cuckoo catfish (*Synodontis multipunctatus*) couple swim onto the scene. They too want to lay their eggs in the nest.[9]

These pale-yellow fish with black spots are smaller than the cichlids they are now making a beeline for. They have long feelers on their faces that look almost like moustaches, and with their flat undersides, they're clearly adapted to swimming close to the bottom, finding food between rocks and grains of sand. They will happily eat a cichlid egg or two, but they're also seeking a lodging for their own eggs.

The cuckoo catfish do the same as the mouth-hatching cichlids: they lay eggs and spray sperm into the nest the male cichlid has made, stopping to sample a few of the nest-builders' eggs into the bargain. The cichlids are visibly distressed. They try to chase the cuckoo catfish away, and at the same time, the female rushes to gather all her eggs into her mouth. And now it's easy to see why this catfish is named after the cuckoo. Without noticing it, the cichlid has picked up a few of the cuckoo catfish's eggs too. Now it won't be her own offspring she harbours: unbeknownst to her, in a few days a bloodbath will take place inside her. For now, the cichlid female conceals herself to incubate the eggs in her mouth, while the cuckoo catfish couple swim off on the hunt for more food and more cichlids to parasitise.

Just as *Ctenochromis horei* uses its mouth both as an incubation chamber and for eating, my uterus has other functions beside housing a foetus. Its primary job is to protect me. Before we had access to elective abortions, paid parental leave and nurseries, the body developed a defence mechanism against us having children we didn't have the capacity to raise. Whereas a bird can leave its nest and its unhatched eggs if it doesn't think they will survive, we have the thick endometrium of the uterus – it protects us against non-viable embryos. It's less resource-intensive to build up the uterine wall only to break it down, than it is to be the stronghold of a foetus for nine months.

As the walls are falling, we eat breakfast. It's a little more than three years since I gave birth for the first time, and the product of this life-changing event spills milk on the table just as she's insisting she can manage by herself. Her dad cleans up, the dog helps with the drips that fall to the floor. The reproductive labour continues for many years after the birth. Our mornings now have an established routine. I'm woken, a little earlier that I would like to be, by the tousled head of a child. As the city around us is gradually waking up, I've already tricked the three-year-old into putting on enough warm clothes for a whole day at nursery without a fight, taken the dog out for a quick leg stretch in the back yard, drunk a coffee and perhaps even managed to glance at the paper, not to mention eating breakfast, mopping up spills, averting tantrums and stressing over whether we'll be able to get out of the door on time for once, leaving the kitchen in complete disarray knowing full well we'll suffer the consequences in eight hours, *and* feeling the muscle cells in my uterus clench to push out the blood.

Menstruation is not particularly practical. It hasn't arisen because it's an advantageous trait, like the opposable thumbs that give us the ability to grip. We don't even know quite why we menstruate, but there are several hypotheses. One is that menstruation is a side effect of the uterus's preparations for the arrival of the egg – like a kind of defence mechanism against an invasive embryo we don't have the resources to support right now.[10] After all, it's not all joy and symbiosis when a fertilised egg attaches to the uterus and grows into a fully formed child. Throughout history, work, starvation, epidemics and existing children have demanded much of those who have uteruses. The body doesn't always have the surplus energy and resources needed to support a growing foetus. The embryo wants to grow as big and strong as possible; it wants everything it can get, but there's no certainty the body wants to give it everything it requires. This hypothesis is called the maternal–foetal conflict. We'll soon see why.

Another hypothesis about why we menstruate and why the uterus prepares itself for the egg's arrival, is that these preparations enable the body to ensure the fertilised egg is viable. If cell division has occurred correctly in the egg and sperm cells, and they have joined properly, the little clump of cells will be dividing flat out to create a new human. It now has to prove its worth by burrowing into the thick lining the uterus has built up to receive it. If it passes this test of strength, perhaps it is also strong enough to become a baby that can live outside the uterus. There's much that can go wrong in the egg's fragile beginnings. Up to a fifth of all known pregnancies lead to miscarriage, but the number of fertilised eggs that fail to become viable foetuses could be up to 50 per cent[11] – because many don't know their egg has been fertilised, and therefore don't realise that what seems like their normal period could actually be a miscarriage. It could be the case that the mucous membranes save the body from wasting huge amounts of energy constructing an embryo that is in some way unviable.

Because if it's not going to make it, it would be just as well to throw the baby out with the bathwater, so to speak.

I don't ask my uterus if this month's egg is worth holding on to, and I can't weigh up the pros and cons before choosing to bleed or become pregnant. It's the process of evolution that has made me, a few other apes, some bats and the little, long-snouted elephant shrew, creatures who bleed.[12] So why are we the only mammals that have periods? A trait that has evolved several times on the tree of life is presumably advantageous for the species that exhibit it. The human embryo, and probably the embryos of the few other species that menstruate, is extremely invasive. It blasts its way in, rather than knocking sweetly; it demands control over our bodies and free access to nutrients. But this is not the case for all species.

On the way out of the door I hold my three-year-old's hand. I have my bag on my back, and she has hers, bouncing along on the way to nursery. We hurry across our habitat, which stretches from the door of our building to the nursery, from the supermarket where we catch our food, all the way to my job in town. The sea anemone, meanwhile, has suctioned itself to a rock in the ocean. It doesn't need to move, so growing a baby from the side of its body is no problem for it.

After just two days the cuckoo catfish eggs hatch inside the cichlid's mouth. Her own eggs, which are also incubating in there, need a little longer. This is fatal for them, because the young cuckoo catfish are making good use of their time. At the very start, their own eggs provide them with a packed lunch – the yolk sac, probably best known to us as the nutritious yellow yolk in the egg of a hen. For the first few days after most fish have hatched, they live primarily off the nutrients the sac provides while they grow large enough to find food for themselves. So too the cuckoo catfish. But five days later, when the cichlid eggs finally hatch, the catfish are ready for more. They soon start eating the little cichlid babies, which the cichlid mother thought safe inside her mouth. In the end, only the catfish are left – the cichlid has fallen hook, line and sinker for a cuckoo that lays its eggs in others' nests.[13]

Thousands of kilometres from this warm African lake, the emperor penguin (*Aptenodytes forsteri*) has finally reached its breeding ground in the middle of the Antarctic ice sheet. These penguins, the largest now living, are as tall as a human four-year-old, about a hundred and ten centimetres, but significantly heavier at thirty-five to forty kilos. The emperor penguin is the only penguin that lays eggs during the Antarctic winter. In order to reach a safe breeding ground where the ice will stay compact until the young are ready to head for the ocean, they have to travel a long way, up to two hundred kilometres. As with many other birds, they live in different places throughout the year, but penguins cannot fly and nor can they run: they have to waddle the whole way on their short legs, in long lines, one after the other.

These black-and-white birds with large yellow patches on their throats now find their mates. They dance with one another, each copying the other's movements, getting acquainted so they'll be able to find one another again. It's absolutely essential that they remember who the other is throughout their long incubation period, so that the female can find her way back to the male and the chick he's now tasked with keeping warm until it hatches. She lays her egg, but in the biting cold things can quickly go wrong. She must get it up on her feet, into the protection of her feathers: it mustn't get cold. She's exhausted from producing the egg, which contains everything the chick will need, and now it's the male's turn to do his bit for the next generation. The egg is transferred carefully onto his feet without it falling on the ground, and he covers it with his feathers – it will lie right next to his skin in a featherless patch between his legs that perfectly encloses it. She's done the first part of the job, and she now returns to the sea.[14][15] He stays behind, and must stand quite still without dropping the egg throughout the cold winter. It seems like madness to lay eggs at the coldest time of year, and there's a great danger he might let the egg fall in a moment of inattention. If that happens, the embryo will immediately die – the cold is brutal. However, winter is also a time when there are no predatory birds about: they've headed northwards, to warmer climes. And in two months' time, when the eggs hatch and the chicks are ready to set off towards the edge of the ice, spring will have come, and with it, food. It takes a lot of effort, but it works.[16]

I wave goodbye to my three-year-old, who's already busy in the sand pit, then I run home and grab my bike. I sail down the hills, through the city streets, and pull up outside my office. The summer holidays have just finished, and the mornings haven't yet started getting cold, so I don't need a jacket over my shirt. I text my partner while I wait for the lift, reassuring him that drop-off went well today. He'll be the one doing pick-up – we take turns with caring duties, keeping each other updated in the meantime, thinking all the while about our child, constantly drawn back to the family unit. Is the penguin mother similarly worried about how her egg is doing while she's away? Would she rather be there to watch over it? What is it that makes her stay away so long, and yet still come back?

Week 2

The flow of blood has dried up, my uterus is done ridding itself of blood and mucus. From the outside I seem completely normal, but within me, my uterus, ovaries and a little part of my brain called the pituitary gland are all hard at work pumping hormones out into my body. Like confetti over a stage, the hormones scatter through my blood stream. My body is doing everything it can to prepare for new life, because an egg is about to be released on its journey to the sperm, to my uterus, to life.

I make dinner and we try to discuss what's been happening in the news lately, to have the kind of adult conversations we used to have, but we soon give up as our three-year-old first requests, then demands, that we read the book she's holding, down on the floor. When we finally sit at the table, she doesn't want to eat; she wants to tell us about dinosaurs, about how a velociraptor had sharp claws, and about how she's been arguing with someone at nursery over who gets to play with the toy triceratops. We're not allowed to talk about grown-up things, we have to tell her what our favourite dinosaurs are and listen as she lists all hers.

*

At the same time, somewhere in North America, a Virginia opossum (*Didelphis virginiana*) has already been through a whole pregnancy. This white-faced, cat-sized mammal, with its long, hairless tail, is found across large parts of the USA. It climbs trees and roofs and tips over rubbish bins in search of food. It might look cute, carrying its gaggle of pint-sized young on its back – right up until it opens its mouth and screeches, baring its sharp teeth.

Just twelve days ago the Virginia opossum mother was tempted by a male who caught her attention by making clicking sounds with his tongue. They mated: he stuck his bifurcated penis into the opening in her abdomen that's used both for reproduction and urination, where the two parts each found one of her two vaginas, and the sperm swam up into each of her two separate uteruses, each of which has its own uterine tube. Afterwards, the two opossums went their separate ways, and she got on with her life alone. He's no longer in the picture – she'll finish the job herself. She starts by licking her fur to create a trail of spittle from her belly up to her pouch, to lead her young to her teats once they're born.

Her hairless, almost transparent young are no larger than a grain of rice, and already they must undertake the most important journey of their lives. They're born quickly through a medial birth canal, a third passage formed between the two vaginas. There's plenty of space in the birth canal and there's no risk of shoulders getting stuck. The little rice-grain babies crawl up to the round hole that leads to their mother's pouch, without any help from her, swimming through her fur using their front legs and temporary claws. These claws are made only for this journey, and afterwards they will fall off and be replaced by permanent nails. It all happens quickly – the babies reach the pouch in two to four minutes – but it's a race to the finish. Inside the pouch there are thirteen nipples, and each little newborn opossum attaches to one as it arrives. On average, the opossum gives birth to sixteen young. Those who don't find a nipple before they've all been taken will eventually fall out and die.[17] There's no neonatal ward, and some won't make it.

In Australia the tiny little sea star *Cryptasterina hystera* is ready. She's not going to carry her eggs around anymore, and nor will she carry her foetuses in a pouch. She is also a he – a hermaphrodite – and doesn't do what most sea stars do, that is, release eggs and sperm into the water. Instead, *Cryptasterina hystera* deposits sperm into its own reproductive opening, the gonopore, fertilising the eggs internally. In the ovotestis, a mix of ovary and testicle that produces both sperm cells and eggs, the eggs grow and divide, becoming larvae that swim around inside, before they are born as tiny sea stars after approximately two weeks.[18] [19] *Cryptasterina hystera* is probably a new species in evolutionary terms, as young as six thousand years old,[20] making it a great example of how the development of foetuses inside the body has evolved multiple times throughout history, and will continue to do so. It's not something only we mammals do.

At long last the three-year-old eats a broccoli stalk and a few pasta twists, muttering on about dinosaurs all the while. Her voice is a continuous soundtrack to our attempts at conversation – sometimes in the background, sometimes so loud and insistent we hear nothing else. She tells us the toy triceratops at nursery is the mummy, because it's the biggest, and all the other animals are its babies, because they're little: the lion and the sheep and the velociraptor are all the triceratops's babies. She pushes back her chair, runs out into the living room, shouting that we have to come, demanding our care and attention, taking the lead role in our little family, the reproductive unit we've chosen to become. She calls again: she wants to build a train track.

Did the corners of the Virginia opossum mum's mouth twitch slyly at those clicking sounds? Did she have butterflies in her stomach when she saw him? Did she feel a desire to mate, or was it just instinct that made her put her tail to one side so the deed could be done? As humans, we know that sex can be enjoyable. In contrast to many other animals, we have sex both when we can get pregnant and when we can't. But we're not the only ones who do so. Bonobo apes (*Pan paniscus*) use sex for conflict resolution and to strengthen social bonds within the troop.[21] They probably wouldn't if they didn't find it pleasurable. Female dolphins (Delphinidae) stroke one another's clitorises, and they have sex even when they're not fertile.[22] Genes want to be spread and sexual urges ensure that they are – but in a surprising number of us, sex is not only something that occurs when it can produce offspring. The Virginia opossum did it voluntarily, she was on heat, she invited him in. If she hadn't wanted it, she would have run away rather than following the clicking sounds. She would simply have gone elsewhere.

It's not my turn to do bedtime today, and after taking the dog out for a walk, I lie down on the sofa while they're quarrelling over pyjamas and teeth-brushing and how many books they're going to read. We've chosen each other, built a nest together, our apartment is our base in the world. Now we want to grow our family by one, we want more wakeful nights and more quarrels as to who's going to brush whose teeth, more childish kisses full of love and snot.

The bed bug (*Cimex lectularius*) cannot choose; she can't run away. These red-brown, five-to-six-millimetre-long insects live in human nests – in our houses – and feed on our blood. They bite us in the night while we sleep, before creeping back through cracks in our walls, beds and sofas.[23] But they don't only bite us. The males bite the females to spread their sperm. The male bed bug has a hard, needle-shaped organ that it uses to transfer sperm. It doesn't do it in the flirtatious, agreeable way other kinds of insects might. The male hoverfly (Syrphidae) courts the female by hovering completely still in the air above her,[24] and some male springtails (Collembola) dance in front of the female, bumping their heads against her until she accepts (or is so charmed that she wants) the sperm he neatly places on the leaf alongside her for her to pick up.[25]

When at last the three-year-old has fallen asleep, we watch a film together, sitting close together on the sofa. I hold his hand, stroking the hairs on his arm, slowly inviting the post-film outcome I'm hoping for.

The bed bug has no time to dance, build relationships, deposit sperm and wait patiently for the female to pick it up. Both the male and the female want to reproduce, and both want to decide how it happens. He's in fierce competition with other males, and wants to make sure it's his sperm that will fertilise the egg, that he will win the race. And she in turn wants to decide whose sperm will fertilise her precious eggs and when they will do so. She's evolved sharp spikes in her genital opening to defend herself against importunate males and help her retain that choice. And yet evolution has given the bed bug male a trick to sneak around those spikes: he uses his hard penis needle to puncture the female's exoskeleton. This enables him to bypass the stage of navigating the needle in between the spikes in her genital opening, something that could really injure him if she changed her mind during the process. He wants to spread his sperm, regardless of what she wants, and so he squirts them into her. They will find their own way through her body to the eggs they are to fertilise, regardless of what organ they're squirted into.

But there's no evolution without co-evolution. The female has evolved a countermeasure: a thickening of the exoskeleton on her abdomen – the area generally punctured by the male – and within it, a pouch filled with immune cells that ensure she can rapidly heal her body after the male has subjected her to what scientists call "traumatic insemination".[26]

*

For me, this time it's both fun and serious. Even as I was menstruating last week, a little gland in my brain had started readying my body for a new egg, a new chance. The pituitary gland produces a hormone with the rather descriptive name FSH – follicle-stimulating hormone. The egg cells that have been stored in my ovaries ever since I myself was a foetus are surrounded by a layer of cells that can produce the hormone oestrogen. The egg cell and its covering of these cells is called a follicle. And when the pituitary gland produces a follicle-stimulating hormone, the follicles start to grow, and the more the follicles grow, the more oestrogen they produce.[27] The oestrogen makes the mucous membranes that line my uterus grow, and so here we go again. Hormones are pulsing through my body, from my brain to my ovaries and from my ovaries to my uterus, and the mature egg is getting ready. Is this the egg that will meet a sperm cell, let it in and perhaps end up as a child?

Suddenly the egg is released. Now it's in free fall along the uterine tube towards my uterus, and it has approximately one day to meet a sperm cell.[28] It's now or never for this free-diving egg, and this time everything works in its favour. I have my male close at hand. We move from the sofa to the bed. The condoms are left in the darkness of the bedside table. The sperm cells swim for their lives and crash into the egg, which selects one candidate for admittance.

Deep in the ocean, ideally in tropical and sub-tropical waters at depths of more than five hundred metres, lives a female fish who has no need to cultivate coupledom once it's been established. A relative of the broad-jawed monkfish, this deep-sea anglerfish, known as the triplewart seadevil (*Cryptopsaras couesii*), resembles nothing so much as a cross between a hairless Persian cat and a bulldozer with a fishing rod, thanks in part to the long growth sprouting from the top of its head. On the tip of this growth, which is a modified dorsal fin bone, there is a colony of bioluminescent bacteria that glow in the dark waters where the sunlight from the sea's surface doesn't reach. But the light is not there to help the triplewart seadevil find its way in the darkness, it's there to lure in prey so she can gobble them up when they get close enough. Her mouth is vertical when closed, her front almost flat – she looks like she's crashed into the side of a mountain. She's between twenty and thirty centimetres in length, and on the side of her body, towards her tail, she has a little growth of around two centimetres. It could be a pimple, but it isn't, it's a

male of the same species. He's bound himself to her and he'll never be free again. He found her when she reached sexual maturity several years ago. Back then he used his large eyes to find her, and she guided him with the bioluminescent bacteria that live not only on her fishing rod, but also on the underside of her body. He sank his teeth into her body and held on, and since then he's gradually lost his eyes, which degenerated as his skin melded with hers. He is dependent on the nourishment he gets from her – he can no longer catch his own food. The only things about him that are still well-developed are his testicles. He's reduced to being an add-on, a vestigial genital.[29] He's a parasite, just as my fertilised egg will become a parasite in my body. But where he was once free, before attaching himself for good to the female's body, my egg will attach itself and be free again in nine months' time.

The world's largest living lizard, the Komodo dragon (*Varanus komodoensis*), is not only known for being so big it can swallow whole goats in one gulp. It has also surprised us by producing young entirely on its own. This two-and-a-half-metre-long animal, with its long, forked tongue, lives only on a few Indonesian islands. It can weigh up to a hundred kilos, and in addition to goats, eats pigs and deer, and even people who come too close, little suspecting this apparently docile, sleepy animal sunning itself in the sand, can suddenly explode in a flash of speed and ferocity.

When she finds a partner, the female Komodo dragon will mate, then lay her eggs. But she has a trick, in case there are no males nearby. She can lay viable eggs that hatch and release little

Komodo dragon babies without ever being fertilised by a male. She has no need of sperm to produce young, she manages it all on her own, through a process known as parthenogenesis. When an egg and a sperm cell fuse, they each contribute their own half of the chromosomes – the structures that contain our genes. In the vast majority of species, an egg cell that hasn't been fertilised cannot develop into a live adult individual; we need the right number of chromosomes to function.[30] When the Komodo dragon lays viable eggs without mating, she herself gives her eggs two copies of the chromosomes. This can happen either by the egg doubling its number of chromosomes or through two eggs fusing together.[31] The result is a viable egg that has received all its genes from the mother. They're still not clones, however: the heritable material has been reordered slightly, and all the eggs the Komodo dragon lays as a result of parthenogenesis are male.[32] It's smart for an island-dwelling species to be able to reproduce alone, for at least two reasons. A female who comes alone to a new island, if carried there by a storm, for example, can lay eggs and start a new colony with her own sons. And if all the males die in a period of food shortage – a likely scenario, given that they have larger bodies and therefore higher energy requirements than females – she can make new males herself, thereby creating someone to mate with.[33]

On an Indonesian island, between herds of grazing deer and idyllic sandy beaches fringed with palms, a Komodo dragon guards the eggs she laid two weeks ago. The hollow she has dug, which may be up to two metres deep, contains about twenty eggs, and for the next fourteen to fifteen weeks she'll stay close to the nest, before suddenly leaving it, long before the eggs hatch around seven months after being laid.[34] The little embryos inside the eggs take a long time to grow – their cells dividing, making forked tongues and feet with claws – but halfway through the brooding period, she stops guarding the nest. Perhaps her maternal instinct wanes after a while, perhaps she gets so emaciated that she leaves in search of food, perhaps there's suddenly no danger of anyone eating the eggs because the rainy season is starting and there's plenty of other sustenance to be found.

I lie in bed, listening to my partner doing the washing up in the kitchen. The bright August night is trying to find its way in through the dark curtains. I would never have wanted to go to bed so early before, but I know I'll have to be up early tomorrow to provide food, warmth and care for my offspring, who still has no idea she's going to be a big sister, that our little family is about to grow.

In Australia the brush-turkey (*Alectura lathami*), a black bird with a bald, bright red head, has also finished laying her eggs for the year. But she doesn't need to watch over them. Though these big birds are called "brush-turkeys" in Australia, and look a little like regular turkeys, the two are not particularly closely related. Their tails are big like a turkey's, but are flattened sideways, like a fan. The bird is a poor flyer, using its wings only to get up into trees at night, or to flee predators.

Over the last few weeks, the brush-turkey has stayed close

to her chosen male's nest, which, to outsiders, looks like a rather large pile of leaves. They've mated, and she has laid several eggs on his mound, at the same time as several other females. His head has been a little redder lately, and the bright yellow wattle around his throat, like a misplaced coxcomb, has grown even larger. This may have impressed the ladies, but his nest is the thing he's really proud of. It's a great big mound that he has painstakingly built up from leaves, mud and other things he found on the ground. As with a compost heap, the mound heats up as the leaves begin to break down, and he ensures the temperature stays at a relatively constant 33°C. He roots around in his mound, adding new leaves, being careful to fill up any holes the females might make as they lay their eggs. When the female feels she has laid enough eggs in one male's nest, she finds another male and lays eggs in his.[35] She's spreading her bets, rather than putting all her eggs in one basket. He takes over and ensures they stay warm enough. He'll do this for seven weeks, but as soon as they hatch, the chicks will have to fend for themselves – they'll get no instructions on how life should be lived, how nests are to be built, or where eggs are to be laid.[36]

Week 3

I'm pregnant, but I don't yet know it.

Normally, it's the female who provides the eggs, either becoming pregnant and carrying the growing embryos in their body, or by laying them straight away. But the reverse is true of seahorses. The male has what's called a brood pouch, and his hottest move is inflating it with water to show potential seahorse mums that it's a good spot for growing seahorse foetuses.

Seahorses swim around like an upright S, with tiny, almost transparent dorsal fins, a long, stiff snout and a ridged tail. They may not look like it, but they are a fish, belonging to the pipefish family (*Syngnathidae*), and they feed predominantly on small crustaceans that they suck in through their long snout. Seahorses are often found in tropical regions, but two of the fifty or so species are found in European waters. White's seahorse (*Hippocampus whitei*) is a species endemic to the Southwest Pacific. It can grow to up to thirteen centimetres in length and is often coloured shades of light brown to black, though some individuals are completely yellow. The male swims around shallow coastal waters, wrapping his tail around seaweed, corals and other things he can latch onto, and flirting with the ladies. He stays within an area of approximately one square metre, while she travels much more widely, until she finds a male she likes and starts courting him. They swim along together holding tails, before latching onto the same piece of seaweed and spinning around it, collaborating to create an interplay of changing colours.[37] When the female's eggs mature, the couple stop moving and join snouts, and she discharges the eggs. For a long time it was thought that she laid them directly in his brood pouch, and that he then released sperm internally into the pouch, in a way that mirrors how human eggs never have to leave the uterine tube and the uterus in order to be fertilised. But there is no internal channel through which he can release sperm into his brood pouch. Evolution has made it possible for him to become pregnant, but his sperm still have to leave his body before they find their way to the eggs. It is likely that the male releases sperm while they are dancing cheek-to-cheek, at the same time as the female discharges her eggs. The sperm mix with the eggs, and in some way we don't yet understand, they find their way into his pouch together, where they can grow into little seahorse babies safe from the dangers all around them.[38]

The seahorse dad incubates the eggs for three weeks. His belly grows and he moves about less and less. It's hard to carry offspring. The pouch does not just provide protection from the rest of the ocean: blood flows via a placenta-like connection, nourishing the growing eggs, and the seahorse dad ensures the levels of salt and oxygen are just right for the foetuses.[39] Every day the female swims into his territory, entwining her tail with his, and they take a little tour after spiralling awhile around a piece of seaweed.

When it's time for the male to give birth, he latches onto a seaweed strand. The contractions shudder right through his body, and the tiny seahorse babies are shot out of his brood pouch – up to two hundred can be born at a time. The father's job is done. The young are out of the pouch, in open waters, and can sail off into the sunset.

As the male seahorse shoots out his young, the egg inside me is implanting in the lining of my uterus. It's millions of years since we came up out of the sea, and it's a long time since female mammals first started to carry their eggs inside their bodies. I'm bound by evolution – I can't just give my eggs to a male.

In my egg, which, now it's embedded, is called an embryo, cell division progresses quickly. The outermost layer of cells around the growing clump bursts into the uterine lining as the cells divide, and begins to form the placenta, the organ that will bind us together and forcibly relocate nutrients from my body to the embryo. The cells from this layer force their way in and enclose my blood vessels. Simultaneously, the budding placenta starts producing a hormone called hCG, or human chorionic gonadotropin. Now things get complicated, but the whole purpose of hCG is to let my body know I'm pregnant. Because, if my body didn't know an embryo had burrowed into my uterine lining, my menstrual cycle would just continue as usual. And then, in just a week, I'd have menstrual blood to deal with once again.

There are still a few follicle cells left in the part of my ovary that released the mature egg. It is here that something called the corpus luteum (Latin for "yellow body") forms. This temporary organ is created each menstrual cycle from the remains of the follicle and produces the hormones progesterone and oestrogen. When the budding placenta produces hCG, this tells the corpus luteum to continue to produce progesterone, thereby stopping the mucous membranes lining the uterus being broken down. If a fertilised egg does not meet the mucous membrane and form a placenta that can produce hCG, the corpus luteum regresses and stops producing progesterone,[40] and the menstrual cycle continues. But this time my body is steered off its regular course: the embryo takes over. The concentration of hCG increases with each passing day. Soon, this hormone will make it clear what's hiding in my uterus.

In thirty-seven weeks I will ecstatically send a picture of myself to friends and family, looking like a serene Madonna figure with a newborn at my breast. But, right now, the relationship between me and the embryo is anything but serene. I'm not a self-sacrificing mother who will do anything for this little bundle of cells, and the embryo is not just some helpless little creature. There's a war being waged, and hCG is a part of that war.

The egg is often described as embedding itself in the lining of the uterus, and it is indeed correct to say that it embeds itself in the endometrium my body has built up. But it's a major understatement to use that word to describe how the embryo's cells burst into the uterine tissue, taking over my blood vessels and building a completely new organ in order to take nutrients away from me.[41]

Before the egg was fertilised, when the uterus was laying down its mucous membrane, something happened to the blood vessels that transport blood to the inner surface of the external wall of the uterus. The outermost blood vessels – those closest to the uterine cavity – are called spiral arteries. This name derives from the fact that when an egg is released, they elongate, curling like spirals or corkscrews. They provide blood for the thickening mucous membrane lining; the spiral form and the muscle cells in the arteries' thick walls enable greater control of the blood flow, allowing the lining to detach and be carried out when the time comes, while ensuring that the flow stops so I don't bleed to death. When the embryo embeds itself and forces its way into the endometrium, it sends out cells towards these spiral arteries. Their thick walls are broken down and built up again, straighter and thinner. Now my body can no longer contract these blood vessels and control how much blood flows through them.[42] The embryo has changed the game. This is a major conflict. The embryo wants to have as much control as possible over access to resources, to ensure that it gets what it needs, while my body wants to make sure that I only give those nutrients to a viable embryo, and that it doesn't take so much from me that I'm lost in the process.

My little embryo is burrowing its way towards my bloodstream. Over the next few weeks, my arteries will start to bring blood right up to the foetal membrane, and the little person-to-be will be able to draw energy and nutrients from my body.

Throughout the history of evolution there has been an arms race between the foetus and the mother, because, as the Australian geneticist and evolutionary biologist David Haig explained through his maternal–foetal conflict theory, the two individuals don't necessarily have the same needs. The foetus wants to live: the combination of alleles[43] it carries exists only here and now, in this individual. And how much energy the mother expends on the foetus inside her will determine whether it comes into the world alive and has a chance of survival. Therefore, it wants to get all the nutrients it can. For the mother, the situation is a little different. Her genes are contained in the foetus she's currently carrying, but they're also in the children she already has, and will be in any foetuses she may carry in the future. Therefore, she stands to gain by not giving the foetus everything it wants; by controlling how much energy she gives that foetus and how much she saves for herself. Perhaps she needs energy to breastfeed a two-year-old now, perhaps her body needs to hold something back so she's not so exhausted after the birth that she's unable to recover, and thus unable to reproduce again. The foetus demands more and more of the mother's resources, and throughout evolution human foetuses have forced their way in, assumed control of the spiral arteries, and taken as much as they can. Alongside this, the mother's genes have evolved to try to restrict the amount of resources the foetus receives to a level that is optimal for the mother and all her offspring.[44]

Foetuses in species that do not menstruate have a placenta that more or less tentatively asks if they might please have a little nourishment from the uterine lining. But with us, the foetus has tried to force itself further and further in towards the nourishing blood, and the mother has responded with thicker and thicker mucous membranes. If a fertilised egg is not strong enough to embed itself and force its way in through the defences our bodies have devised, then our bodies will simply bleed out the membrane, and along with it, the egg.

It's one of the last warm summer evenings of the year. We're sitting on the balcony, and I take two sips of my partner's beer. It might be the last taste of alcohol I get for a long time. The few drops taste good, forbidden. How the next year will turn out, whether I'll be drinking beer or growing a baby, depends entirely on how the war taking place within the silence of my uterus plays out.

Week 4

I should have had my period by now, but I know why it hasn't arrived. When I got up this morning, I managed to interest the three-year-old in a sticker book so I could go to the toilet in peace for once. I weed in a cup, grabbed the rectangular white box on the top shelf of the bathroom cabinet, and wrestled the test out of its plastic wrapper. I took off the lid, dipped the test in the urine, put the lid back on and put the test on the kitchen counter before going back and pouring the rest of the cup's contents down the toilet, and all before her enthusiasm for stickers had lapsed. As I filled a bowl with porridge and milk and bargained over the necessary quantity of jam, the level of hCG in my urine made two blue lines gradually appear on the test stick.

I'm no longer alone in my body. I'm a breeding apparatus, an incubator, but not for real yet. It will be many weeks before I can say it out loud; anything can happen with Schrödinger's embryo,[45] it's alive and taking over my body, but it might stop living at any point, tumbling out in a clump of blood and mucus. I hope I'll be carrying it inside me over the months to come.

The female common eider (*Somateria mollissima*) sits on her nest for four weeks. She can't head out to sea to catch food. If she did, the icy wind would immediately kill the embryos inside her eggs, so she can only leave the nest occasionally for short periods to drink water. She gets thinner and thinner, but she puts up with it. As soon as the ducklings make cracks in their shells with the temporary egg tooth on their bills, and stick their downy little heads out, she'll take them down to the water. By the time the eggs finally hatch, she might have lost up to 40 per cent of her body mass. She has to constantly weigh up whether she should stay on the nest or abandon the eggs to save herself – to find food so she can survive to potentially raise young next year instead.

She's lined the nest with her soft down, the down that is the reason she became the first bird in the Nordic region to be domesticated, long before chickens, mallards and Muscovy ducks took over. She's large, speckled brown in colour, and has a distinctive bill that forms a straight line from her brow to its tip. Along the coasts of Norway, Iceland and the Faroe Islands, people have for centuries given the female eider (the duck) reasons to make her nest close to human dwellings.[46] While the male (the drake) stays out at sea and joins the other males in large flocks to moult their feathers, the duck heads for land, generally to the same nesting site year after year. And there are some very safe places for her to nest: places where people have built little sheds for her and prepared seaweed as nesting material and will stop cats and dogs from bothering her throughout her stay. The difficult brooding period is made less trying in proximity to humans, who keep predators at bay. In return, we get to collect the down from inside her nest when she and her young have left it. It will be used to make the world's best – and most expensive – duvets. A duvet filled with Norwegian eider down can cost more than five thousand pounds and might help keep people warm for several generations.[47]

When the female eider leaves the nest with her freshly hatched ducklings, she's only completed the first part of her job. There's a few weeks of intensive care in the shallows awaiting her before her young are big enough to follow her out to deep water where they can dive for the best food. But by now the duck is hungry and exhausted. If she's too worn out to look after her young in shallow water, she'll leave them with another female. Up to half of eider ducks make this choice after finding someone else to look after their young. There's better food to be found in deep water, where the little ones can't yet go. Perhaps leaving them behind is the only chance of survival she – and they –have.

Are those who take care of others' ducklings simply being nice? Do they run these large-scale nurseries out of the kindness of their hearts? The mothers who work together in this way gain clear advantages from it – they can be away from their young a little longer, dive deeper for better food. They take it in turns to watch for gulls who might like to snack on a tasty little eider duckling. But why do some eider ducks watch over abandoned ducklings even when the mothers have left for good? There may be a sound explanation for it. The eider ducklings find their own food, rather than getting it directly from the adults, so it's not necessarily a problem to have a few more of them trailing after you. Perhaps the seemingly altruistic ducks make a simple probability assessment: the more ducklings there are, the less chance there is of one's own young being carried off by a hungry gull. It could be an advantage to have a few young with you that you don't care too much about. When an eider duck cries out a warning about an incoming predatory bird, her biological young react faster, swimming right in under her wings. It's the adopted young who are at greatest risk of being eaten, acting as "gull fodder" to help ensure the safety of her own offspring.[48]

For the duck on her nest, it all comes down to a precise evolutionary calculation: Am I capable of raising this brood, or will I die trying? Should I save my strength for next year, for the next clutch? She's already weighed up whether she would lay eggs at all this year, or whether she should use the summer to fatten herself up and build her strength. And the eider duck is not unique in this – she's just one of many animals that have to make decisions about timing where pregnancy is concerned.

When the brown bear (*Ursus arctos*) becomes pregnant, the eggs do not immediately embed in the lining of her uterus. Instead, after they've divided a few times and become little clumps of cells, the fertilised eggs pause their development before entering the uterus. The clumps of cells are kept at this stage for several months, until the mother is securely hibernating in her lair. But not even now can the embryos grow freely. If the mother hasn't put on sufficient fat stores to get through a winter of pregnancy and suckling in hibernation, the fertilised eggs do not embed.[49][50] The mother doesn't waste energy on cubs she won't have the strength to raise. Instead, she cuts her losses and waits for better times.

Humans, too, have a mechanism by which we avoid getting pregnant if we don't have enough energy for our long pregnancy. Though we don't utilise embryonic diapause like the brown bear, if we don't have enough fat on our bodies we simply don't ovulate.[51] This is a smart strategy in times of famine, if we're undernourished, or if circumstances are otherwise out of kilter.

There's no famine in my life. I've ovulated and I'm pregnant. I can carry my belly about, I can eat what I need to, I don't have to sit and keep my brood warm. This embryo has got its claws into me, it wants to get somewhere in life. Soon it's going to put me out of action for several months: I'm going to start feeling nauseous.

The eggs of the platypus (*Ornithorhynchus anatinus*) hatch after four weeks, just like those of the common eider, but they spend almost three of those weeks inside their mother's body. She is an egg-laying mammal, her otter-like body complete with a broad tail and a large flat bill. She lives in the water and lays eggs, but she also provides her young with milk; almost like us, yet different. While her eggs are still inside her body, she can move about and seek food, but for the last ten days, when they are outside her body but not yet hatched, she lies quite still in her burrow with them on her stomach, protected by the long tail she's folded up towards her bill.

Soon the young, or "puggles", as they're known, will break through the soft shell of their eggs with the little tooth on their bills, which then falls out just as a bird's egg tooth does, or the first claws of baby Virginia opossums. The mother is no longer gestating – her body is free – but her young are naked, bald and completely dependent on her. She gives them milk, but she doesn't have teats – the milk runs out into her fur from small pores on her stomach. The puggles lap it up with their little bills, and don't leave the burrow until they are three or four months old. After a while the mother is able to go out in search of food, staying away for longer and longer periods. She needs sustenance to produce milk, and every day she must eat a huge quantity of insects, shrimp and crabs – almost equivalent to her own bodyweight, and four to five times more than she would normally eat.⁵² ⁵³ By way of comparison, a breast-feeding human female needs to increase her food intake by approximately a quarter, but it's still thirty-seven weeks until I'm at that stage. My body doesn't yet need anything extra, because my embryo is so small.

We play in the back yard after nursery. The three-year-old wants to look for insects, so we pick up rocks and dig in the flowerbeds. I lift a loose slab, and the woodlice that have made their home beneath it run for cover. We catch some of them, flipping them over to see if they have eggs hidden between their legs.

The common pill woodlouse (*Armadillidium vulgare*), an eighteen-millimetre-long trilobite-like animal that can be found under rocks across much of Europe, is not an insect. Woodlice are crustaceans that have come up onto land – like our ancestors, they emerged from the water to make use of new grazing grounds, to find new habitats. They are in the isopod group on the evolutionary tree, an order of crustaceans that includes more than ten thousand species in the sea, in fresh water, and on land. The woodlouse still breathes through gills, so it's dependent on damp places, even if the hard shell on its upper body does a good job of holding in bodily fluids and stopping the sun drying it out. The woodlouse has seven pairs of legs – a clear indicator that it is neither insect nor spider (they have three or four pairs, respectively) – between which the woodlouse mother conceals her young in brood pouches until they're old enough to be born. She carries her young in the pouches for four long weeks. They lie there, bathed in fluid, growing so large that, towards the end, the mother is incapable of using her most important defence mechanism: rolling herself into a ball – her belly quite simply gets in the way. She's weighed down by the means of reproduction that evolution happens to have given her, just as I will be in nine months' time, losing the ability to tie my own shoelaces. Once this batch of young are finally out and her brood pouch is empty, the common pill woodlouse will have more broods – she can live for many years and produce many more offspring.[54]

Week 5

The little cluster of cells inside my uterus keeps dividing. As long as I don't start bleeding, I can assume they're alive. The cells begin to specialise: at this point the cluster has a top and a bottom, a front and a back, and the vague hint of what will be a spine.[55] The downsides of pregnancy have begun for me: the morning sickness makes itself felt, as if to confirm the embryo's presence. A parasite has got its claws into me, my body has been taken over. It begins as faint discomfort, but I've felt it before and I know what it leads to. I might as well tell my boss right away – I have a steady job, and the law ensures that I can't be fired just because I'm reproducing. I don't need to hide it – I need to prepare my employer for the possibility that I may soon be absent for many weeks. But my uncertainty is growing. Is this really the best time for me to get pregnant, with this particular child?

After five weeks, the rabbit (*Oryctolagus cuniculus*) gives birth to its small, hairless young. Apart from a lack of fur, which will soon grow in, they're fully formed, with long ears on top and a soft belly underneath, a snout in front and a tail behind. They lie perfectly still in the nest their mother has made for them from dry grass and her own soft fur. To avoid marking it with her scent and thereby attracting predators, she only visits the nest once or twice a day to suckle them. Does the mother rabbit think of her young while she's away from them, finding enough food for herself so she'll be able to produce the milk they need? Does she wonder if a predator has found its way to the nest, if their little hearts are still beating as they nestle inside the burrow?

I think about my little cluster of cells almost all the time. Are they still dividing? Do I have any symptoms of being pregnant? Do I have any symptoms of *not* being pregnant? It's many weeks until the hospital calls me in for an ultrasound and I can actually see if there's something in there, not just air and hope. For now I have to trust my body, and I ask the bartender to give me alcohol-free beer, but in a glass so I can avoid any questions about why I'm not drinking. Meanwhile, I go on planning the autumn like I'm not going to be spending it in bed, clutching a sick bucket.

A Namaqua chameleon (*Chamaeleo namaquensis*) is ready to lay her eggs. For five weeks she's been carrying them inside her, but now she will push them out, bury them and leave them behind. It's still fifteen weeks until the foetuses will be able to clamber out of the eggs, but from now on they'll no longer be protected by her body, but by the moist sand down in the hole she's dug for them.

The Namaqua chameleon is a desert species, and unlike most other chameleons, she stays on the ground, running across the warm sand, patrolling her territory, catching insects and small lizards by shooting out her long tongue, and chasing away other chameleons if it's not mating season. Five weeks ago she mated with a male whose territory lies nearby. After the initial encounter everything happened pretty quickly. At first they stared at each other. He wiggled his head and body to show he wanted to mate, and she considered whether she should chase him away or accept him. She's bigger than him, so it's easy for her to decide what will happen, but this time she decided to mate. He inserted one of his hemipenises into her cloaca and fertilised her eggs before she chased him out of her territory again. Like most scaled reptiles, or "squamata", the chameleon has only one abdominal opening – the cloaca – and this takes care of both reproduction and waste excretion. The male's genitals are concealed within his cloaca. They are in two parts (hence the "hemi", meaning half), and can swell up, so they are rather reminiscent of the penis we're familiar with from our own species. The difference is that they often have barbs or hooks with which to hold themselves inside the female, and that one testicle is connected to each hemipenis. This means he doesn't use up all his sperm at once: he has a reserve in case another female comes along before he can replenish the stores he's used.[56]

The next task for the female is digging her hole in the sand, either by deepening her sleeping hollow, or excavating elsewhere in her territory. She has to get down to moist earth to ensure her eggs won't dry out. When she's happy with the hole, she deposits her eggs in layers, first one or two, then shovelling sand over them before laying a few more, up to about ten. At last she covers the hole over with yet more sand. Now she has to eat and rest a little to build up her strength. Then she starts her work again. While the first eggs are incubating, she lays a new batch, depositing two to three clutches in the course of the season.[57]

As my embryo is burrowing into the lining of my uterus, the foetus of the eastern grey kangaroo (*Macropus giganteus*) is crawling up her abdomen. The mother kangaroo is leaning her body backwards, putting her weight on her tail and her heels, and pushing up her vaginal opening to make it easy for the hairless, bean-sized foetus to crawl up into her pouch.[58] But is it really still a foetus, if it's been born? It doesn't have any fur, its hind legs are under-developed, its eyelids are still fused together, and its whole body will change over the coming months in the place it's currently heading for: the pouch. This little bean is going to latch on to a teat, which will swell up in its mouth so that it is unable to lose its grip. In fact, the teat will become a kind of umbilical cord. The foetus is outside its mother's body, but is still protected by it, receiving nourishment in the pouch the mother has cleaned and prepared ahead of the birth. Her joey has overcome its life's first journey, and it will stay here for a long time – eleven months – and remain dependent on its mother's milk for a further nine months after that.[59]

Meanwhile, the mother has entered oestrus again. She mates, and a new egg is fertilised. She is multitasking: she has a baby in her pouch, a fertilised egg on its way to the uterus, and a joey at her side who is too big to be in her pouch, but still gets a little milk from the teat not occupied by the bean-baby. The fertilised egg will not keep growing for the time being. It will remain a clump of cells, ready to develop further, but it must wait while the baby in her pouch is attached to the teat. When it is ready to face the world and is able to jump out, sticking its head in now and then for a little milk, the development of the new foetus will proceed. It will go on growing for a further five weeks until the time comes for *its* journey up to the pouch.[60]

So, just like the brown bear, the kangaroo can put its next pregnancy on hold. The embryo patiently waits its turn, and doesn't start developing until the mother's body signals that there's a space free. This means she has a new baby in the pipeline should she lose the one in her pouch when fleeing a forest fire or if food is scarce. Or perhaps, if she's lucky, it will grow big enough to fend for itself. Then she can have the next one straight away, without needing to wait for a male to come along and provide sperm.

In the case of the bear and many other animals, the embryonic diapause lasts for a fixed amount of time and happens regardless. This makes it possible to use the autumn to fatten oneself up, rather than using one's resources to find a partner. But with the kangaroo, the diapause is dependent on circumstances: the embryo is kept on ice, so to speak, and starts developing the moment there is no joey nursing in the pouch. It's an adaptation that allows the female to respond to the possibility of losing young without having to immediately find a male.[61]

*

We are copying the kangaroo when we store our fertilised eggs in a fertility clinic: this technological diapause means that we can implant embryos one by one, crossing our fingers that they will attach to a uterus that, for a variety of reasons, might need help. I didn't need help, but I would love to be able to pause the embryo at the cell-ball stage. I've been pregnant before and I know my body doesn't handle it well. I've experienced morning sickness and I'd like to wait a while before doing so again. I've planned this pregnancy, tried to become pregnant and succeeded, but I'm ambivalent. I'd like to know that I'm pregnant, feel safe in the knowledge that the egg has embedded, and then be able to pause the process. I'd like to bank this embryo so I can finish the exciting project I've got on at work before letting a new life take over my body. But I can't plan exactly when I get pregnant, I can only try and then be glad when it finally happens. That means the timing won't necessarily be perfect. I would rather have vomited my way through the dark days of winter than these sunny ones full of beautiful autumn colours. I would have liked to attend that conference with work in a few weeks' time. But I can't: the embryo has other ideas.

Week 6

Nausea floods my body. It crashes over me in waves, and I'm left hugging the toilet bowl. I stagger back to my room. I have a bucket by my bed, and I nibble dry crackers and take slow, slow sips of water. My body is growing weak, and my mind can do nothing but focus on getting my body through it. From one day to the next I go from being a rational person with good ideas in my morning meeting, witty rejoinders over lunch and time to meet a friend for coffee, to a mere receptacle, a shell around an embryo, existing solely to ensure that it survives.

At first, I vomit as soon as I get up and then function normally for the rest of the day. But suddenly one morning I realise that it's begun. It's too late. I throw up once, then again. I hope it's going to pass, that a short rest on the sofa before going to work will do the trick, but then I'm sick again. And again. I manage to let work know I'm not coming in, manage to tell my partner I'm feeling unwell, then my rational brain switches off entirely.

Now all that matters is getting from the bed to the bathroom and back. Little sips of water come straight back up.

Coffee is out of the question. I go from being a highly capable person to nothing but a body with needs. My system is out of balance and everything comes back up again – and it's this alien organism that's making it happen.

The internet tries to reassure me that it will pass, but I know that this is not necessarily true. Some of us get such strong nausea that we stop functioning, and our embryos make us more and more lethargic while they go on getting bigger and stronger. I have to stay in the warm, dark cave of my room, cut off from the world that's rushing by out there. The three-year-old skips around me, her every movement making me more nauseous. My partner cautiously caters to the ever more specific food requests that trickle through to him via text message over the course of the day: lightly toasted bread with butter and thinly sliced cheese. French fries with no seasoning. Thin slices of those expensive Pink Lady apples.

My embryo has attached itself to me in earnest – it wants nourishment from my body, and that's making me sick. Even though it's only three millimetres long and looks like a funny little worm, it's stopping me from being able to do the simplest things, like eating an ordinary meal, taking the dog for a walk or reading to my three-year-old.

It must be some kind of evolutionary mistake, that I get so ill. It feels like I wouldn't have survived if it weren't for the bland meals my partner brings me in bed, the anti-nausea drugs my doctor prescribes me, the sick note that enables me to lie in the dark all day while the three-year-old is at nursery and my colleagues send pictures from the annual conference, which is taking place on a boat this year. Would I have survived in a stone-age cave, or as one of the very first humans two hundred thousand years ago?

The leading scientific hypothesis is that morning sickness exists to protect both the mother and the embryo.[62] The symptoms are at their most virulent in the weeks when the little grub in my belly is most at risk of being harmed by the dangerous substances I might decide to eat or drink. During this time, I am also at risk of serious illness from contaminated food. My immune system is at a low ebb to ensure that my body doesn't forcibly eject my uterus's new occupant.

Much is still unknown about morning sickness. Why do only some people get so ill they can't properly care for themselves? You might think the embryo inside me has extra vitality and is pumping loads of hormones into my blood, making me more nauseous. But there's no difference in hormone levels in the bodies of women who do get sick and those who don't. My only consolation is that there's a very slightly lower risk of miscarriage in women who are actually sick than for those who are only nauseous. But it really doesn't feel like that's worth all the vomiting.

The most interesting thing about studies into morning sickness may be the fact that there are some indigenous cultures that report no incidence of nausea during pregnancy. It seems that across the world there are populations where pregnant women do not throw up, where they don't spend their days regurgitating their last meal or navigating a world where everything is constantly spinning. These communities live on predominantly plant-based diets, and often on diets consisting largely of maize. But there are also communities subsisting on plant-based, maize-heavy diets in which the women do suffer from morning sickness, so it's clearly not just a case of eating maize throughout your life.

Avoiding animal products is quite a smart thing for the gestating body to do, and the foods pregnant women around the world tend to feel most nauseous thinking about are meat, fish and eggs. Meat stored at room temperature – both raw and prepared – offers a good breeding ground for bacteria and moulds.[63] A study has shown that it is possible to predict whether a community will experience cases of morning sickness by looking at whether their diet features low intake of cereals and high intake of meat, sugar, oilseed crops and alcohol.[64] Unfortunately for me, statistics are compiled at the population level. I drink relatively little alcohol and don't eat meat, but I'm still out for the count, incapable of doing anything but turning my stomach inside out.

Do other species get morning sickness? Does a guinea pig mother-to-be lie in her burrow wanting nothing but fresh blades of grass, and does anyone come to her aid? Vomiting during pregnancy is only documented in humans, and there are very few known examples of diet changes in other animals. One is that dogs tend to eat less in weeks three to five of their nine-week pregnancies. Rhesus macaques in captivity also tend to have lower appetites in weeks three to five of the twenty-three weeks they are pregnant, and this is followed by hormonal changes comparable to those in the first trimester of human pregnancies. Only once has there been a documented case of chimpanzees in captivity experiencing morning sickness, but it hasn't been reported in any other studies of this species – one that has been studied quite extensively. Why don't our closest relatives get sick like us?

If morning sickness is an adaptive phenomenon, something that has evolved because it offers us an advantage, there are good reasons why humans are the only species to experience it. Humans have an enormously broad diet compared with most other mammals, including other apes. This has enabled us to spread to every one of the world's continents, surviving on the food we have been able to catch, find or farm there. But it also has consequences. Evolving enzymes that can break down everything in the food we eat is resource intensive, and there are some toxins we haven't yet got the measure of. Perhaps the nausea, vomiting and accompanying aversion to certain kinds of food is actually a defence mechanism.[65]

In any case, it's misleading to call it morning sickness, as I can testify. Pregnant people are nauseous both evening and night, and if they're particularly unlucky, they can vomit around the clock.[66] A more neutral term, and one that means these side-effects cannot be dismissed with the argument that "pregnancy is not a sickness" is "Nausea and vomiting in pregnancy" (NVP).

I won't die, even if I can't cook my own food or look after myself these next few months. I'm a member of a social species, and we collaborate when it comes to reproducing, finding food, building dwellings and raising children. We help each other. And we're not the only ones.

The eggs of a little spider with the rather descriptive name the African social spider (*Stegodyphus dumicola*) are hatching at about this time.[67] These centimetre-long, hairy arachnids weave large, dense, communal nests of silk around branches and leaves. Food is caught in separate webs nearby. The nest is used over several generations, and there can be anything from dozens to thousands of spiders in residence. They live in communities with a high degree of inbreeding. And this might be why the females are prepared to sacrifice everything for the young in their group, despite the fact that many of them have not laid eggs themselves.

Only females are left to look after the eggs now that they're hatching. The males die just a few weeks after reaching adulthood, while many of the females take longer than the males to reach sexual maturity. That's why up to 60 per cent of the females who are poised to look after the young haven't laid any eggs themselves. All the same, they've worked to prepare for the eggs that are about to hatch, six weeks after they were laid. They've helped to build and pass along the sacs that contain the eggs, they've guarded the nest from intruders and they've contributed food to the group. Now, as the eggs hatch and the hairy little young climb out, the help takes a new form. Both the females who have laid eggs, and those who weren't old enough, begin to feed the young with insects and a kind of fluid they produce from their own bodies.[68] This has fatal consequences. Over the coming months, the adult spiders will regurgitate digested food, allowing the young to feed straight from their mouths. But it's not only food the adults offer up – it's also their own bodies. In a month's time, the first female will die from feeding the young on her own innards, which she has coughed up for them. And when she does, the baby spiders will suck out the rest of the fluid from her body until there's nothing left but a dry husk. The eating of a mother by her offspring is called matriphagy, but here, it's not just the mother they are eating. Over the coming months, most of the females will be eaten by the young, acting both as protectors and living packed lunches for the next generation, until the brood are big enough to look after themselves and it's their turn to eat and be eaten.

The spider aunties who haven't had progeny of their own are much more closely related to the young who end up eating them than human aunts and uncles are to their nieces and nephews. Those females who reach the sperm too late, and are unable to mate, can't take off and try their luck elsewhere. The best option they have for passing on some of their genes are the young who are waiting to hatch, even if they are not their own offspring. They are not necessarily self-sacrificing altruists, keen to do anything for baby spiders regardless of their parentage; it may just be the only chance they have to contribute their genetic make-up to the future of evolution.[69]

Week 7

The nausea is a darkness that envelopes me. I float silently in and out of the waves it sends through my body, through my mind. I feel sick if I listen to the radio, if I look at a screen, if I think. I lie in my bed in the dark. I start the day by vomiting, I end the day by vomiting. I put all my strength into a brief chat with my three-year-old after nursery – she comes into the room, wanting a cuddle, her sudden movements making me nauseous. I'm afraid she might put pressure on my stomach; if she does I'll projectile vomit all over her. She gives me a hug before she goes to bed, and having her chubby little arms around my throat makes me gag. It feels as though I'm never going to get out of this hole, out into the world again. My body is the vessel for a growing embryo, and with every cell division, it gets stronger while I get weaker.

There is a group of fungi – Cordyceps – that parasitise the bodies of their host insects. The fungal spores infiltrate the insect – a caterpillar, for example – and slowly consume it from within. The fungus lives off the caterpillar's body while it's still alive, controlling its brain to make it crawl to places that are advantageous to the fungus, whether that means burrowing into the ground or climbing up a blade of grass swaying in the wind. In the end, the caterpillar dies, leaving nothing but an empty husk full of fungus, which now bursts out of its host, forming a club-shaped growth that pokes out of the caterpillar's head and spreads new spores that will infect new caterpillars.[70] To an extent, I can relate. I am controlled by an embryo, and in the end it will burst out of me: not from my head, but through my vagina. At least I won't be eaten up, even if it feels that way right now.

A text pings in from a friend who wants me to come and work out with her, like we usually do. I reply that I can't, letting her in on the secret of my parasite. She tells me that she too is pregnant – just a week or so further down the line than me. She's never felt better: she's working out, going to work, her skin is glowing. I'm happy for her, of course, but I can't bring myself to respond – it's such an isolating feeling: being glad I managed to get pregnant, but also unbearably sick as a result of my pregnancy. I'm preoccupied by the prospect of the months I'll spend lying here in the dark; congratulating others is beyond me right now.

I still have time to change my mind, to get my body back, to dodge the work that lies ahead. In Norway, we have access to elective abortions up to twelve weeks, a right we've fought hard for, giving women the power to decide over their own bodies, even if an embryo has taken up residence. But this time I want to have this baby, and it feels like that's much more important than my personal well-being over the months to come. I want to feel this baby lying on top of me. I want to embrace it, to put it to my breast, to protect it from the world. I want to have this baby much more than anything else right now. I offer up my body to this little creature – it's going to have power over me for the next few months, and it's decided that I have to stay in bed.

Why do I even want this? I could have had a nice life without another round of sleepless nights, early mornings and dirty nappies. I like being able to talk to other adults without being interrupted, to read the newspaper in peace and spend time on my hobbies, just like many others. Some of us have children, some choose not to, some are unable to have children and feel a great loss. The question is perhaps not so much why I in particular suddenly want to have kids, but what evolution really is.

The underlying evolutionary instinct in all individuals is to reproduce, to pass on one's genes. Reproduction is the currency of evolution.[71] Because if there hadn't been an instinctive drive in all individuals – to survive, to eat rather than be eaten, to reproduce – then we wouldn't exist. From the primordial soup to the current day, every one of your ancestors, in a direct line from the first single-celled organisms, reproduced. Reproduction started with simple cell division more than 3.5 million years ago, accompanied our journey up out of the sea and then into the trees, then down to the ground once more.

Your body is the product of all those who passed on their genes.

But just because each and every individual who led to you being here right now – your parents, grandparents, great-grandparents, all your ancestors for the last 3.5 million years – survived and reproduced, that doesn't mean you must do the same. Countless individuals have died without passing on their genes, mostly owing to circumstances or because they got an allele or mutation that made them easy pickings for death. But free will also plays a part: evolution has given us humans a brain with the ability to question reproduction, to choose other things in life than having children.

Evolution is not a force that depends on thought. It isn't guided, it doesn't follow a particular direction. Evolution is the fact of a population changing in genetic terms, which can, in time, lead to new types of organisms. This occurs through natural selection,[72] a phenomenon that means some individuals survive and have more children than others because they are carriers of alleles that are advantageous in the precise time and place in which they find themselves.

Is there a biological clock ticking in the background that makes me feel I have to have this baby, or is it the society around me influencing me into believing that the three-year-old ought to have a sibling? Or is it the welfare state and the promise of affordable childcare? Does the eastern grey kangaroo consider whether to have another now or wait a while? Does the sea anemone wonder how cloning will work in the future? Would the eider duck have preferred to do something else with its time? Does the platypus have a hobby she'd planned to get into before finding out she was pregnant by that hot young male she just couldn't resist – does she regret her life choices now she has all these eggs to brood?

Week 8

A giant Pacific octopus (*Enteroctopus dofleini*) lies completely still in her cave. She has stuck her eggs to the ceiling, and they hang down in great clumps, hundreds of thousands of them. She measures four metres from the tip of one tentacle to the tip of the opposite one: she's as long as an estate car, and when she crawled into the cave around eight weeks ago, she weighed as much as fifty kilos. She hasn't eaten since then, hasn't ventured out, hasn't done anything but guard her eggs. The only movement she makes is to flush water over them with her enormous syphon, the tube she previously used to shoot water through when propelling herself through the ocean. She's making sure the eggs get enough oxygen, and that parasites and algae don't get attached to them. With her sensitive suckers, she strokes the eggs, examining them carefully and checking that even those in the middle of the clumps get fresh water. She is slowly starving, willing to sacrifice everything for her eggs, for her young. She will stay here until they hatch in four or five months' time, depending on the water temperature.[73] [74] Does she get hungry? Does her digestive system complain when she doesn't give her body nourishment? We have no way of knowing, but we will return to her in week 22, when her myriad offspring are ready to embark on their own journeys under the sea.

I'm still in my own cave, my dark bedroom. I'm an octopus whose only aim is to look after my offspring, the embryo inside me. I vomit to ensure no parasite comes near my egg. I run my hands carefully over my belly. I'll be lying here a long time. I should be glad there's a new person growing inside me, grateful it's gone well so far, but I can't. Being nauseous is not like having a little pain in my foot, something I can distance myself from, dull with painkillers, and push through. It undermines everything I try to do, and no matter how hard I fight it, it won't go away. I can't think when I'm this sick, I can't care for the three-year-old when I'm vomiting, I can't make conversation with my partner when I'm anticipating my stomach's next attempt to turn itself inside out. I can only lie perfectly still and hope it passes.

Week 9

When I was little, I had guinea pigs (*Cavia porcellus*). They weren't meant to have babies, because we had two females... didn't we? Then one day we realised Pernille, with the tufts of brown and white fur, was really a boy. It felt like an eternity from the moment when we realised the other guinea pig was pregnant until she gave birth. Gradually she got fatter and fatter, her stomach got huge, and she went from running around to waddling. Right now, the nine weeks of a guinea pig's pregnancy feel like the blink of an eye compared with how long I still have left. At this stage, you can't even see I'm pregnant by looking at my body, and she was already giving birth to a litter of fully developed young. Almost as soon as they have come into the world, guinea pig babies are able to run away from danger. They're born with fur, open eyes, and little humanesque, hairless ears. They will get milk from their mother for a few more weeks, but they're already beginning to find their own food as well.

My little embryo, which from this point on is called a foetus, weighs only a few grams, but it's growing quickly – it's two centimetres long at the beginning of the week and three by the end. Its heart is beating, and the outer ear is beginning to look the way we expect human ears to look, but the eyelids are still fused together. It did have gills, but they're no longer present and the gill arches are in the process of forming the structures of the inner ear.[75] At this stage it's far from being able to find its own food or survive by itself; it will be living off me for many months to come.

Even though guinea pigs are tiny and walk on four legs, and we're large and walk on two, our skeletons are constructed in the same way. We have roughly the same bones in the same places – ribs, shoulder blades, spines and pelvises. The pelvis is the bone that forms a ring between the hind legs and the spine in all those who evolved from a common four-legged ancestor. Before the guinea pig gets pregnant, and in the early stages of her pregnancy, the diameter of her pelvis is eleven millimetres. A newborn guinea pig's head has a diameter of twenty millimetres, almost twice as big. Something has to happen to her pelvis before it's time for her young to come out – they must pass through that ring, there's no other way. In both humans and guinea pigs, the two sides of the pelvis meet at the front, along a joint called the symphysis, which is filled with a disc of cartilage. A ligament stretches from one bone to the other to hold the symphysis together. During pregnancy, the symphysis stretches, and the guinea pig's pelvis increases to up to twenty-three millimetres in diameter. The same hormone that expands the guinea pig's pelvis also expands ours, but whereas the human pelvis expands by a few millimetres across the symphysis – with all the pain that involves – the guinea pig's will expand to the point that its diameter doubles. If it didn't, the young would never get out.[76]

But they did, it worked, they're living, breathing and pretty much fending for themselves. And their mother can mate immediately after giving birth and have a new litter after a further nine weeks. It feels ridiculous to even think about having sex so soon. The only thing I wanted at that point last time was to be able to pee without a catheter, to get up from a chair without feeling like my uterus was about to fall out through my vagina, and to look after my baby. My goal was to make it to the next breastfeed, the next shower, the next opportunity to lie down and close my eyes. Whether or not the female guinea pig chooses to mate again right now, at this moment in time, her pelvis and the skin covering her abdomen are ready to contract, while my pelvis hasn't even begun to move. My foetus is the size of an olive, my belly hasn't grown, and I'm still lying in bed with my eyes closed, because it reduces the risk of me throwing up. I pitch around in my dark bedroom, my very own version of a uterus – my phone the umbilical cord connecting me to the world. Both my foetus and I have our eyes closed.

In Antarctica the male emperor penguin has been incubating his egg for nine weeks of snowstorms and freezing temperatures. In the huge brooding colony, the males huddle together to protect themselves from the cold, changing position now and then so that those who were in the centre of the group take their turn at the edge, with their backs to the wind to keep in the warmth. Eggs that fall from their father's feet crack straight away in the cold, which is fatal to the little bird foetus inside. But our male has managed to keep the egg on his toes, close to his skin. He's tired – the egg hatched a few days ago, and he's been feeding the fluffy little hatchling with a milk-like substance he produces in his stomach. Now he has no reserves left, either for himself or the chick, so it's mum's turn. Luckily, she comes back, her stomach full of krill and

fish to raise the baby on. She hasn't been eaten by a leopard seal or an orca, and has she realised she needs to achieve her true potential instead of raising children, and when she's done waddling the long way back inland to the brooding site, she finds a ready-hatched chick.

Now it's her turn to look after her offspring while the male returns to the water to fatten himself up, having lost up to half his body weight. In three to four weeks he'll come back, and it will be his turn again, and they will rotate in this way right up until the chick can look after itself, six months after the egg was laid.[77] I can't share my incubation period, but when my chick has hatched, I too will be able to take turns with my partner. Over the years to come, we'll rock, comfort and carry our kids, play with them, teach them, provide food and clothing, and gradually see them grow more and more independent, before we let them leave us to live their own lives, waiting for them to come back again when suddenly it's us needing help and care from them.

Week 10

The kiwi is not like other birds. It can't fly, its feathers resemble fur, and it bends forward as it walks, with its weight at the front of its body and its long beak to the ground. It looks like it's about to fall on its face. But its feet, with their long toes and claws, help to balance it. Kiwis spend their lives on the ground in the rainforests of New Zealand. As there were no predators in the place they evolved, wings were unnecessary. Their nostrils are almost at the tips of their beaks, in contrast to most birds, whose nostrils are close to their heads. But this forward position makes sense when you consider the fact that they are nocturnal and the way they scurry around their dark rainforest territory sniffing out earthworms and insects. Kiwis are often monogamous – they live together in pairs – and the little spotted kiwi (*Apteryx owenii*), has been observed cohabiting with its mate for as long as ten years.

This pint-sized bird, small for a kiwi species at around a kilo in weight, is found in only a few locations in New Zealand. To indicate that they are paired, the male and the female dance up close, crossing their beaks and using them to point at the ground, then move in circles, grunting softly, for up to twenty minutes. Their performance confirms that they're a couple, just as my partner confirms that he will stay here and look after me by bringing me toast and slices of Jarlsberg as I lie in my darkened room while my body makes him a baby.

When the little spotted kiwis are preparing to reproduce, the male digs a hole in the ground, into which the female will lay her large egg, which weighs as much as a quarter of her body weight. She has packed an enormous lunch for her little foetus, with the egg consisting of as much as 60 per cent yolk, much more than that of other birds.[78] The little foetus needs all it can get, because it's going to spend a long time in there before it's big enough to live outside its protective calcium shell.

The kiwi female has expended a lot of resources in making the egg, eating large quantities of food in order to produce the nutritious yolk. All in all, it takes thirty days, but that period is not counted when we talk about the exceptionally long time required to make a baby kiwi. Only the brooding time is considered: the ten weeks once the egg is laid before the chick finally emerges into the world, which is a very long time compared to most bird species. The reason that her efforts in producing the egg are not included – even though it takes four weeks and she has to eat three times as much as normal during that period – may be that it has proved difficult to find out how long she spends making it without cutting up lots of birds to look at their ovaries, something that would be foolish to do with a rare species. Or perhaps it's because ornithologists have traditionally been men, and the fact of a female bird having an egg inside her was seen as so ordinary that it didn't seem relevant when calculating who made the greatest reproductive contribution.

Either way, by the time the egg is ready to be laid, the female is worn out. Her belly is bulging so much that it drags on the ground, and carrying her heavy load means walking with her legs wide apart. Female kiwis have been observed resting in cold water, perhaps to soothe the stretched skin on their abdomens and lighten the load on their muscles and their bodies as a whole by temporarily relieving the weight. In the final days before the egg is laid – her equivalent of giving birth – she's unable to eat. The egg takes up so much space in her body that there's no room for food.[79]

Even when the egg is finally laid, the little spotted kiwi chick still has a long way to go before it's ready to emerge. Its countdown starts now, after its mother has made this huge effort. In several kiwi species, the parents divide the brooding period, but with the little spotted kiwi, the male does the incubating alone, and is seen by humans observers as an extremely self-sacrificing father, while the female somehow gets less credit for managing to lay an egg that amounts to a quarter of her bodyweight. He sits on the nest while she stays nearby looking after him. After all, they are a pair who will stay together for many years yet. Is her cloaca torn when she's finished laying the egg? Does it feel as though her insides are being rearranged again when it finally comes out?

To help them carry and push out the enormous eggs, kiwis have a stiff tail, an enlarged ribcage and a very curved pelvis.

This pelvic shape, unusual in a bird, allows her to carry the egg beneath her pelvis so that it does not restrict the egg's size. The kiwi also has a special way of walking: instead of increasing the pace of their steps when they want to go faster, they simply take longer strides. This peculiar gait is probably an evolutionary compromise between being able to move about and having to carry such a large egg.[80]

Every day, the male leaves the egg to go and eat. Now and then he covers the nest hollow with twigs and leaves. It wasn't so necessary before invaders from Europe arrived, and he hasn't yet realised that he now needs to do it every time. Previously, kiwis had no enemies, they could scamper around the forest floor without being eaten, and they could leave their eggs without them getting stolen. Then along came the Europeans with their cats and dogs and other animals, putting the male, the female and the egg at risk of being eaten by foreign species.

When the kiwi chick is at last ready to emerge into the world, it has to crush the egg with its feet, since it doesn't have an egg-tooth like other birds. It's fully developed, with scruffy, fur-like feathers, looking just like a little adult. It still has much of its packed lunch, the egg yolk, in its stomach, and can survive on that for a week or so before it starts eating independently. Its parents don't give it food, even though it stays close to the nest for a while before going off to find a territory of its own. After five or six days it starts to move about outside the hollow, and after two to three weeks it's probably completely independent of its parents, even though it might remain in their territory for a while longer.[81]

But why does the kiwi's egg get so large she can barely walk in late pregnancy? Couldn't she just lay it a little earlier, as most birds do? Presumably, it's because, this way, the chick can survive by itself almost as soon as it hatches – it has feathers and it can walk, and it has a belly full of egg yolk that provides it with an initial store of energy. Kiwis previously had no need to fear egg-eating, ground-dwelling predators – the big threat was from other birds who loved to eat the helpless little chicks.[82] It's worth putting the time into making and brooding an exceptionally large egg no-one can get their beak around, with a chick that can look after itself as soon as it hatches – at least until new predators suddenly show up and start gorging themselves on confused kiwis who have no evolutionary defence against that kind of attack.

The Pacific beetle cockroach (*Diploptera punctata*), which lives on Hawaii and other Pacific islands, is also giving birth now, after ten weeks, to between nine and fourteen little "nymphs" that will go through multiple stages before they are fully grown. At no point have they been left in defenceless eggs that might be eaten by predators: they've been protected inside the mother's body, which has also provided them with nourishment. She's given her foetuses a kind of uterine milk rich in carbohydrates and proteins, which sustains them while they're in the brood sac within her body. While she's gestating, running gets increasingly difficult, becoming impossible as the foetuses get bigger. She needs more food, just as I do.[83] It's the price she pays for protecting them with her own body rather than laying eggs externally, as the majority of insects do. We're in the same boat.

Far away from me, far from the kiwi, and the cockroach, somewhere in the rainforests of Valdivia on the west coast of the southern part of South America, the young of the Southern Darwin's frog (*Rhinoderma darwinii*) are finally crawling out of the male's vocal sac.[84] Once the female – brownish-green with a pointed snout – has laid her eggs on the damp ground, the male fertilises them before settling down to guard them for twenty days. When the eggs begin to move, he picks them up in his mouth, but instead of eating them, as he would an insect moving about in front of him, he puts them into his vocal sac, the growth on his throat that he ordinarily uses for croaking. The vocal sac becomes a brood sac, and this is where the tadpoles hatch from the eggs and then live for more than fifty days. They eat the yolks of their eggs, but that's not the only thing on the menu. They are also nourished by a fluid he provides, produced by the wall of his vocal sac. While in the sac, they develop into froglets, going from having gills and tails to lungs and legs. But now their time in his body is over, as is his time carrying them, and he can have his vocal sac back. He can go back to croaking at new females and fertilising new eggs to protect.[85] [86]

Week 11

The wandering albatross (*Diomedea exulans*) is one of the longest brooders in the avian kingdom. This large sea bird, with an enormous wingspan of more than three metres – the largest of any living bird – spends eleven weeks sitting on its nest before the ten-centimetre-long egg hatches. Females lay an egg only every other year, because it takes twelve months to feed up a chick from a hatchling to a juvenile able to fly alone. The parents do this work together, joined in a lifelong monogamous pairing.[87] Now the egg hatches, and the long-winged little chick peeps out, well protected by the warm body of its mother or father.

I'm eleven weeks into my long gestation, and there's still months to go before my young peeps out and shows its little head. I pitch in and out of nausea, vomit, then lie still in the darkness. When the three-year-old comes in to see me in the afternoon, she smells of the cool autumn. Her cheek is cold against mine, her skin fresh against my sweatiness. I curse my ovaries for binding me to this bed. Why am I the one who has to carry this baby? Why am I the one with a uterus?

Inside me, a tiny human is growing. It's a male, with testicles instead of ovaries. Towards the end of this week, it will be possible to see that on an ultrasound, but we won't find out for several weeks. Modern humans tend to categorise each other in two big sex groupings – women and men – based on how our genitals look and whether we make large or small sex cells: eggs or sperm. That hasn't always been the case. Historically there have been many examples of human societies where sex was not only divided into two separate groups, but was understood to have fuzzy edges or to be subdivided into several categories.[88]

As soon as I find out the sex of my foetus, I will involuntarily begin to imagine what they will be like when they grow up. I try as hard as I can not to let gender norms constrain the way I behave towards my three-year-old, but it's difficult in a society that dresses babies in pale blue or pale pink on the basis of how their sex organs look. What is sex anyway, and how does it shape who we are? Is the urge to distinguish so clearly between people on the basis of their genitals rooted in biology, or is it something we've invented as part of our culture? Wouldn't it have been easier just to be hermaphrodites, so we could each make both large and small sex cells? What about having more than twenty-three thousand mating types,[89] like the split-gill mushroom (*Schizophyllum commune*)?

Sex evolved in the last common ancestor of all eukaryotes (organisms with cell nuclei – that is fungi, plants and animals)[90] around 2 billion years ago.[91] The first sex cells were probably all the same size. But gradually, over many generations, they grew large or small. This probably started tentatively, with the odd individual that occasionally produced some sex cells that were a tiny bit larger than normal. These slightly larger sex cells were able to give more nourishment to their growing progeny. This was an advantage that gave it a head-start over its rivals, and therefore something that was carried on through the evolutionary process. But that didn't simply lead to all sex cells getting larger and larger. Some individuals would occasionally generate sex cells that were a tiny bit smaller than normal, with the advantage that a few more cells could be produced overall.[92] This meant that there was a race to generate both bigger and smaller sex cells. Continuing to turn out sex cells of medium size was no longer working, because there was a greater advantage associated with producing either fewer, larger, nutrient-rich sex cells, or a lot of small ones. And that's when what we call females and males arose. Females produce the large sex cells – the eggs – but generate fewer of them. And males produce the small ones – the sperm – in much larger quantities. This is what we're normally talking about when we talk about biological sex.

These two strategies for producing sex cells have had evolutionary effects in the millennia since they first occurred. In most species, we can see a difference between the sex organs of individuals who produce large sex cells and those who produce small ones. If we can't tell from the outside, we can if we cut them open. And the evolutionary effect doesn't just manifest in appearances. Traditionally, people have said that it also has a major impact on behaviour. It has been said that because the female produces only a few, large eggs, and the male many small sperm, the female will have committed more resources to the reproductive process and will therefore be more invested in the sex cells she has produced. And that females of species that have to do more to reproduce than simply releasing their sex cells into the water, or mating and laying their eggs, will be more caring. According to biologists, those females that produce these large eggs – and make a large investment – will want to take extra care in guarding or brooding their eggs, or offering food, protection and knowledge to their offspring until they can manage by themselves. Meanwhile the males, who produce a large number of small sex cells, will be more concerned with spreading their sperm as widely as possible, and will therefore use their time to compete with other males to get access to as many females as they can. The females only mate with one male, while the males mate with many females. This interpretation of the impact of sex cell size on behaviour has followed us ever since we first understood how reproduction works, and now forms the basis of how we see sex and gender roles in human society. It also laid the foundations for why we dress boys in pale blue and girls in pale pink.[93] But when we begin to look more carefully, this understanding breaks down.

Let's take birds as an example. Perhaps you've heard that birds are generally monogamous? Indeed, many bird species have been held up as heteronormative paragons of "true love". In the book *Bitch: A Revolutionary Guide to Sex, Evolution & the Female Animal*, the biologist Lucy Cooke shows how difficult it was to unpick this myth. She explains the opposition faced by professor of evolutionary biology Patricia Gowaty, who used new DNA technology to show that birds are often socially – but not sexually – monogamous: 90 per cent of all female birds mate with more than one male. When Gowaty presented her findings, her male colleagues assumed that it was the male who was the active party in these extra-marital escapades, that he forced himself on the female, and that the females weren't particularly interested. Females were seen as passive, with no interest in mating outside the monogamous relationship they had entered into, while males had other plans altogether.

The problem with this assumption, as Cooke shows in her book, is that only 3 per cent of all bird species have penises. In species where the male isn't thus endowed, the male and female have to collaborate in the fertilisation process by each being willing to open its cloaca – their genital opening – to the other's. Male birds without a penis quite simply cannot force themselves on a female. When one puts aside one's assumptions, it's not so difficult to understand that, from an evolutionary perspective, it can be an advantage for both males and females to mate with more than one partner,[94] in case one of them proves to have less robust genes than the others. For many species, it makes sense to avoid putting all one's eggs in one basket, or all one's sex cells in one nest. Scientists didn't grasp this until they were able to imagine that the world could be different from what they believed. They had to set aside their assumptions in order to see what was happening in the natural world around them.

*

Throughout history we've not only believed that it's males who are the more eager party in any reproductive event, we've also assumed that it's always a male and a female that raise young together in those species where two adults care for the offspring. In some species it's easy to see the difference between the sexes, while in others the male and female look exactly the same from the outside. When scientists began to check the sexes of pair-forming birds in which the sexes are largely indistinguishable, they realised that it was not unusual for two males or two females to be found together, as observed with the king penguin, the Laysan albatross and the grey goose. As soon as you look more closely, you realise that nature is full of species that don't fit the heteronormative narrative of mother, father, child.[95]

Our assumptions about sex, and our inability to imagine anything other than what we expect to see, go a long way back, but were cemented by Charles Darwin. He gave us his theory of evolution a hundred and sixty years ago, fundamentally shaping our understanding of how species change over time. It was revolutionary. He was able to think outside the box, to understand something no-one before him had been able to articulate, to see the world with completely new eyes. Our knowledge of evolution today stems from observations he made when he spent five years travelling the world aboard HMS *Beagle* in the 1830s. In 1859 he published *On the Origin of Species*, his book on how evolution occurs through the process of natural selection. But when, ten years later, he proposed his hypothesis on sexual selection – that is, what determines which individuals get to mate – he wasn't able to see past the narrow gender roles he himself had grown up with.

Darwin was fascinated by why the sexes were so different. Why does the peacock (*Pavo* spp.) have such a large, colourful tail, when the peahen looks frankly rather dull? His answer was that the males are fighting for the chance to mate with the females, and it's the female's role to choose a male. The female contributes by being particularly taken with certain features, in what's known as "female choice", which prompts the development of the slightly odd traits we see in males, like the peacock's enormous tail feathers. But Darwin thought that evolution really occurred in the male of the species, because it is their reproductive success that varies. All females would probably have a chance to mate, while a male's success was dependent on whether or not they were able to compete with their rivals. He thought that it was the males fighting it out who had the competitive instinct, who had passion, while the females had to be courted: they were "coy".

Strutting men and bashful women are gender roles lifted straight out of the English Victorian society Darwin lived in. And he was succeeded by a line of white, male natural historians who described species through the lens of their own rose-and-blue-tinted glasses. It wasn't until women started studying biology that the field was able to view nature with greater objectivity.[96]

And it turned out that females compete too, both for access to food and other resources and by mating with more than one male. Social status in the group or access to territory and food also impacts their reproductive success – it's not only the males who have something to fight for. The female does not simply wait for the best male to come along and mate with her; to ensure her own genes are passed on to the next generation, it makes sense for her to mate with multiple males. And males do not necessarily mate with any old female either. Their sperm does not spring from some infinite source, even if there are more sperm cells than there are eggs. There's actually a limit to how many times a male can release a batch of sperm, and doing so doesn't always lead to him producing offspring.[97]

But isn't it females who provide the most care, who spend the most time guarding eggs and young? Well... Who bears the burden of care after mating varies by species, though generally, when it comes to mammals, it *is* females who do the lion's share of the work. But there are more species of fish in which the males guard eggs and spawn alone than those in which both parents provide care and in which only the females do.[98] And in a large review of bird studies, scientists concluded that there was no relationship between how much resources each of the sexes invested in producing sex cells, and who took on most of the work in caring for eggs and young.[99]

Though it does seem that the females in many groups generally put more energy into looking after progeny, we still don't know why this is. The traditional explanation – that the female has already invested more in her eggs and is therefore more inclined than the male to undertake caring duties – is criticised not only because it ignores the fact that both females and males compete, but also because it assumes that since the female has already used more energy than the male in producing sex cells, she'll have more to lose if she gives them up. This presumes that the female's early investment in the eggs impacts what she chooses to do in the future, rather than her choosing at a later point the best course to optimise her outcomes. This logical pitfall is known as the Concorde fallacy, taking its name from the British and French governments' continued investment in supersonic airliners during the eighties and nineties, after it had become clear it wasn't an economically viable project. They weren't able to pull the plug before they had given their all to the aeroplane.[100] When an eider duck abandons her eggs, she gives up all the energy she's already put into that year's brood, but in doing so, she is prioritising her own survival, allowing her to raise more young in the years to come. If she'd gone on investing in the eggs, like the British and French governments did with Corcorde, it could have ended in death – not only for her eggs, but also for her, and therefore for all her future offspring.

Sex is smart because it creates variation. When the sea anemone clones itself, and a new individual grows right out of its stalk, this new individual is exactly the same as its parent. If there's a fatal genetic trait in the parent sea anemone, it will probably also be present in the offspring. With the development of sex, individuals began to mix their genes, and completely new, unique individuals were created. We're all a little different, because we have some alleles from one parent and some from the other.

Sexual reproduction comes into its own when the environment around individuals changes, as, of course, it often does.

If the climate gets a little warmer, for instance, the individuals who have a genetic make-up that enables them to function a little better in warmer conditions will produce a few more offspring than those whose genetic make-up means they function less well at higher temperatures. In this way, species with sexual reproduction can adapt to changes over time. But the costs associated with sex are still significant: breeding individuals need to invest time in finding each other, and there are only half the number of children per parent when two parents have to share them. The time spent competing for a mate could be used much more efficiently – for example, in the care of fertilised eggs or young. Sex is a waste of energy. It's a biological paradox that sexual reproduction is the dominant form of species propagation, even though asexual reproduction is so much more energy efficient.[101] [102]

Water fleas (Cladocera) are tiny insects that grow to no longer than five millimetres in length. They swim about in lakes and eat small particles they filter out of the water with the arm-like growths at the front of their bodies. When conditions allow, they lay eggs by parthenogenesis, as the Komodo dragon can. They keep the eggs, which don't need to be fertilised, in a brood sac on their backs. In favourable conditions – long, sunny days that are perfect for the growth of the water flea's favourite food, small algae – the female produces offspring completely independently. She tries to fill the lake with her own progeny – to be as prolific as possible and spread her genes – without having to spend time in search of a partner and his sperm. But when the days are shorter or the temperature is lower, or if the lake gets over-populated or nutrient-poor, she changes strategy. That's when she produces eggs that are fertilised by a male – eggs that are not just her own, but a mix of both parents' genes.[103]

Experiments show that if a parasite attacks a population of water fleas produced either sexually or asexually, those who are the offspring of two parents fare much better. They are more tolerant of infections, while the parasite grows much faster in water fleas produced through parthenogenesis.[104] Asexual reproduction is smart when conditions are good, and food and space are plentiful. But if the environment changes, all those genetically similar individuals are in trouble.[105] *E. coli* bacteria might be able to produce countless new offspring in the time it takes me to grow a single human, but sex is present in many species simply because it works.

I still can't shake off the thought that I would rather have avoided this pregnancy. If my partner could have shouldered the burden, could have been the one with the uterus this time, while I delivered the sperm, I would have changed places in a heartbeat to get out of being a walking incubator for another parasite-like foetus. It would have been practical to share this task, to be able to alternate who carries the foetus each time.

And there are plenty of species out there that allow individuals to be both sexes in one body, either simultaneously, like snails and earthworms, or subsequently, like in many fish species. The fish from Disney's *Finding Nemo* – the orange-and-white clownfish (*Amphiprion ocellaris*) – is born as a male, and the sex-shift is triggered socially when there are no females in the population: one of the males in the group changes sex, and becomes the group's female.[106] Being both sexes at once is an advantage when the probability of encountering others of the same species is low – it means you can be sure of mating when you finally do meet someone. And in certain species, single individuals can even mate with themselves if all else fails. But not all the hermaphroditic species we have – most belong to the plant kingdom, but there are many examples in animal groups too – can be explained by a low probability of finding someone to mate with.

In some species, when two hermaphrodites meet, they fight. They each try to be the first to spike the other with their penis, because whoever strikes first gets to be the male, while the one on the receiving end has to be the female, which in these particular species means expending more energy on eggs. But in other species, such as earthworms, the eggs of both parties are fertilised by the other's sperm, and the burden is divided equally.

In those species who are first one sex, then the other – sequential hermaphrodites – it's often advantageous to change sex when one reaches a certain size – because at that point they are large enough to produce many large eggs, or so large that they have access to females.[107] But why are there no mammalian hermaphrodites? Apart from a few fish, there are very few – if any – hermaphroditic species that are large or more advanced. Perhaps that's because reproductive success requires more than just the presence of sex organs. We have to compete with others, invest in our offspring in various ways, and survive long enough to reproduce. Being able to change sex during your lifetime might simply be too costly compared to persevering with the sex cells you have and trying to reproduce that way.

Back to the land of non-hermaphrodites, where some are fated to produce eggs, while others make sperm. It's still an evolutionary mystery why females are predominantly the ones who spend the most time caring for offspring.[108] Biological theorists have struggled to get to the bottom of it. Some scientists have tried to use the same mechanism that so fascinated Darwin to explain the disparity. They advance, as Darwin did, the hypothesis that sexual selection – which includes all the traits that influence an individual's ability to reproduce, and thereby pass on their genes[109] – primarily occurs in males. Let's take as an example the peacock's tail, which we considered earlier. It's enormous, it makes flying difficult, and it makes the males easier prey for predators. But still, over the generations, females have chosen to mate with males with larger and more ornate tails. Perhaps it's because it indicates that he's able to survive in spite of his unnecessary showing off: he must be extraordinarily strong, as well as adept at evading predators. But for males, sexual selection based on their characteristics comes at a cost. Males must plough resources into the characteristics the female favours, leaving them with fewer resources to invest in caring for eggs and young.

If biologists are to use sexual selection to explain why the females of a species are the ones providing care, a number of conditions must be fulfilled. First, some males must fail to reproduce, while others thrive, and there must be a correlation between the traits they put their resources into – their tail, for instance – and their mating success. In addition, the female has to mate with many males, making it difficult for the male to know which young he has fathered, because, if he knew which were his, he would have a greater incentive to care for them. And we would also expect there to be an imbalance of adult females within the population, which can be the result of intense competition to mate leading to high mortality among males. These conditions aren't met in all species, but it's still mostly the female who provides care, and we can't quite explain why.[110]

When we think of sex in humans, we often envisage two types, based on who produces large or small sex cells. Large sex cells mean you're a female, small that you're a male. Every time we meet someone, our brains categorise them as either "woman/girl" or "man/boy". We interpret what we see – hairstyle, clothing and posture – and use it as a basis for determining whether the person has testicles or ovaries. But which sex cells a body produces is not the only way of thinking about biological sex. David Crews, a professor of zoology and psychology, has listed five characteristics that can influence whether we interpret an organism as being a male or a female: which chromosomes it has, which sex organs it has, what hormones its body produces, how its body looks, and how it behaves.[111] Let's take chromosomes in humans as an example. Women have two X chromosomes, men have one X and one Y chromosome. But there are known mutations that can mean an individual has the chromosomes for male sex (XY), but still has female sex organs. Or that they have an extra X chromosome (XXY) and male sex organs but also develop breasts during puberty.

One to two per cent of people have neither the combination "XY chromosomes and male sex organs", nor "XX chromosomes and female sex organs".[112] In addition, there are all those people whose sex chromosomes correspond to their sex organs, but whose identity doesn't match the one they were given at birth based on how their sex organs look.[113] When we're talking about sex, we often act as though nature has made two completely separate boxes for males and females. But it's not that simple. Evolution is variation, both between species and within them. If we didn't have variation, it would have had nothing to work with. We have various basic models of what sex is, and for how different species look, but these models are disrupted all the time. That's the nature of evolution.

The terms "female" and "male" also connote more than which sex cells these individuals produce – just think how we explained the behaviour of male and female birds. We look at the genitals of a human baby and immediately imagine how the individual in question will be when they grow up. If those genitals look like they produce large sex cells, we might expect – whether we want to or not – a child who at the age of three will enjoy playing with dolls, threading beads or doing similar fine-motor activities, and who is kind and thoughtful to those younger than her. If those genitals look like they produce small sex cells, we imagine a child who at the same age will want to play with cars, jump down from high walls and shout and fight quite a lot. We expect those who produce large sex cells to look and conduct themselves in a different way throughout their lives from those who produce small sex cells. This what we mean by gender norms.

Just as the traditional understanding of what large and small sex cells mean for an individual's reproductive contribution is being unpicked – be that caring for young or competing to spread sperm – so too is the understanding of how we view biological sex. When we think of males and females, whether humans or other species, we don't only picture which type of sex cells they produce. We immediately get ideas about how they live their lives – as passive females or active males, and about which feelings and urges are strongest in them: the maternal instinct versus the impulse to compete. But human females can have very high levels of testosterone – which we've traditionally viewed as a male hormone – and human males with XY chromosomes can have female genitals. So what is it that determines our sex?

It was hard for Darwin to look at sex without interpreting it in the light of the cultural context in which he was raised. This is a little odd, given that he was able to think so far out of the box when he made history's most important contribution to biology: the theory of evolution, which was a huge break with the "God the creator" mentality he'd grown up with. Likewise, it's hard for us today to see past what we've learned about sex in nature and what it means for us and our gender roles. And society's expectations of how we will behave has major consequences for how we end up living our lives.

The wandering albatross on its nest may be socially monogamous, but we don't know if she has mated outside her primary relationship, or whether her partner really is a male – it's just an assumption. I know nothing about what interests my foetus will have when he grows up, but the expectations society greets him with will shape him.[114] How we're raised has consequences for the way our brains function later on. It's extremely difficult to know what human sex differences are triggered by: how much is culture – the way we're raised – and how much is biology and genes? Biological divergence is not necessarily as great as we think, and categories bleed into one another depending on what we're focusing on. After all, we're humans first and foremost, not females and males. Throughout history, the divide between the sexes has been stark, and women have been denied education and the right to vote on the basis that it's not in their nature to take an equal part in society alongside men. Today it would be outrageous in the West to suggest that women should be denied the right to vote or get the education they want. Perhaps traditional gender roles will be the next "natural" distinction to fall, as society develops?

Week 12

At last, my first trimester is over. The foetus is real now. I can tell people why I've been absent, put paid to the myths going round at work that I've got *E. coli* or burn-out. The whole time, the foetus has been showing me that it wants to be here, that it's in it for the duration. After the first trimester, the likelihood of miscarriage is low enough that you're traditionally "permitted" to say you're pregnant, as though a miscarriage is something to be ashamed of.

For most, the nausea has passed by now, but not for me. However, I have become a little more human. My brain has been permitted to return to its normal duties, but not to resume full control. Unfortunately, I still can't make it in to work: my body has been hired out and my brain's no longer in the driver's seat.

By the time my foetus is at last officially real, the Nile crocodile (*Crocodylus niloticus*) has finished brooding her eggs. It all started with a drawn-out flirtation in the water, during which the female and the male rubbed their necks against each other's jaws until she accepted him as the sperm donor for her children-to-be. He climbed on top of her and wrapped his tail around hers to bring their cloacae close to one another. The male's penis, which is constantly erect, popped out of his cloaca to guide his sperm into her cloaca.[115] We don't know why crocodile penises are always erect, it's one of the mysteries of evolution – as the clitoris was for a long time. All mammals have clitorises, but they're not the only ones. The female crocodile not only has a clitoris, she has an unusually complex clitoral structure. For a long time, scientists wondered what she might use it for, before drily concluding that it was probably stimulated by the male as they mated.[116] Throughout history they have been more interested in the function of the penis than the clitoris. The two organs are produced from the same structures in the cluster of cells growing inside the mother, but that hasn't stopped people thinking that males in particular need to be driven to reproduce by the lure of sexual pleasure, while females simply take what they're given. Luckily, there's a growing understanding of the fact that the nerves in a female's genitals contribute to sexual arousal, and that females also enjoy pleasure during mating. The experience of pleasure is a biological reward for both females and males,[117] evolved to ensure that we can be bothered to reproduce.

Even though the Nile crocodile lives in the water most of the time, it's descended from a land-dwelling animal and must come back up onto land to lay its eggs: the foetuses inside will die if they are laid underwater. But it's not enough for the nest to be on land. It has to have just the right humidity, just the right temperature, and it has to be somewhere the female can guard it. When the female Nile crocodile is choosing a location for her nest, nothing is left to chance. Her reptilian brain doesn't only go on instinct – she chooses carefully based on previous experience, selecting a site that will offer the greatest probability of her eggs hatching.

She makes her nest in a sandy bank, in an area that won't flood during her long brooding period, close to vegetation in which she can hide and water she can drink.[118] She considers several locations before deciding. If anything unpleasant happened in a location where she previously laid eggs, such as scientists capturing her or predators plundering her nest, she will remember it and may avoid this area in future. It seems that she evaluates locations throughout the year, noting the availability of water through dry and rainy seasons[119] and storing this information for future use. After picking a spot, she digs a hole in the ground, lays her eggs in layers on top of each other, and covers them with sand. Then the waiting starts. For three months she stays close to the nest. She tends to lie on top of it at night and keep to nearby shade during the day, always on land and always close by, even if that's less safe for her than slipping into the water.

Now twelve weeks have passed,[120] and the eggs are about to hatch. They're so deep in the sand that the young can't dig their own way out – without their mother's help, they'll die. When they start squeaking, she immediately gets digging, letting her young emerge into the light. She removes the sand with her front legs and snout, and she can help the young break their eggs by crushing or rolling them in her mouth. She also uses her mouth to carry the hatchlings down to the water, where she will continue to guard them for several weeks.[121] She doesn't eat them – she knows they're hers, their squeaks stir something in her.

The duration of the Nile crocodile's brooding period is determined by how warm the ground is. She has to dig her nest just deep enough – it can't be too cold, or the foetuses will die, and the same applies if it's too hot. But it's not just a case of finding a place that's not *dangerously* warm or cold, because the temperature in the nest will also determine the sex of the little crocodiles. If it's a little on the warm or cool side, the Nile crocodile embryos will develop into females, and if it's somewhere in between, they'll be males.[122] If she wants some of the eggs to turn out male and some to turn out female, which gives her the best chance that some of her young will one day reproduce, she needs to find a place where she can bury them in strata, with those buried deepest becoming females, those in the middle becoming males, and those closest to the surface perhaps becoming females if it's very warm.

But why is sex determined by temperature? Isn't it risky if only one sex is suddenly produced? The species could die out. The question of why evolution has resulted in the sex of so many reptiles being determined by temperature or other environmental factors is still open,[123] [124] but since this trait is found in so many species, it must have worked over time. The species survive and new eggs are laid, becoming new young – new males and females – generation after generation, at least until climate change confuses the nest-digging females so much that all Nile crocodiles end up being the same sex.

Week 13

I spend the day in my dark cave, but I can see daylight: I draw the curtain a little, follow the strip of light that slowly moves across the bedroom wall over the course of the day. Outside, the trees are gradually changing colour. I can look down on the cars passing, the people walking along the pavement, to work, to nursery, to the shops. I'm drawn to the light, I want to go out in it, but my body won't let me.

Somewhere far away from where I lie in the dark lives a mother who doesn't want to come into the light, who will perhaps never see it.

Around this time, down beneath the sands of the Kalahari Desert in southern Africa, the Damaraland mole-rat (*Cryptomys damarensis*) gives birth to her young, each weighing eight to ten grams. They're not rats in the usual sense, like the brown rat (*Rattus norvegicus*), but something that could possibly be described as a mix of mole and rat. So it makes sense that the family of species they belong to is called the mole rats – they, like all their cousins, live underground within intricate systems of burrows.

The Damaraland mole-rat lives in small colonies of between eight and twenty-five individuals who collaborate in burrowing, finding food and raising young. The adult animals are also small, weighing around a hundred and thirty grams, with tiny eyes, barrel-shaped bodies and short legs. Their ears are almost invisible. The only thing that distinguishes their drab little furry brown bodies, apart from a white patch on the head, are the large incisors, which stick up in front of their lips like enormous skewers, emerging from both the upper and lower jaws.

The female who has just given birth in one of the shared sleeping quarters in the colony's great system of burrows – the queen – is the only reproductively active female in the group. She will nurse the young for almost a month, though they will begin to eat solid food after just six days. She's a birthing machine, producing to up to three litters of around six young a year, and is the only one who contributes to the group's expansion. She and the reproductively active male are the leaders, and as with a wolf (Canis lupus) pack, only the senior pair has young. The other members of the Damaraland mole-rat colony find their places beneath them, in a hierarchy based on size. If two individuals can't agree on who is bigger, they will tussle a little, lock teeth to see who is stronger, and then quickly come to an agreement as to who's boss.

Even though only the queen gives birth and nurses, the other mole-rats also play their part. They ensure the young don't stray too far from the nest, carrying them back carefully between their large teeth, they fetch food from their stores and dig passages in search of more, they repair tunnels and living spaces, and they raise the alarm and help to move the young if intruders arrive. Whenever scientists have unearthed a colony to find out what they're actually up to, these Damaraland mole-rats, the largest, reproductively active female is the very last one they catch. The other members of the colony go into battle on her behalf, sacrificing themselves to protect the one who produces new life for the group.

In the areas where the Damaraland mole-rat lives, many plants hoard their resources underground. When it rains, they rush to store water and nutrients for the long dry spells inside huge swollen roots or stalks, just as potato plants do with their tubers. The mole-rats use this to their advantage: just as we eat the potato, they eat these tubers. They live their whole lives underground, burrowing their way from tuber to tuber, depositing them in storage rooms and relocating their sleeping quarters, living quarters and separate toilet burrows to the areas where they have the largest stores. They only go up to the surface to deposit earth from the long passages they dig. They hurriedly shovel it up out of a temporary hole before sealing the opening to stop snakes coming down into the passages. The only time the mole-rats go above ground is when they head out into the world to start a new colony. They don't mate with family members, so every now and then a young mole rat full of hope will set off to find a mate, so they can be the one to reproduce.

Damaraland mole-rats do not have large eyes, and the areas of the brain that process optical information have shrunk. In the dark, they have no need to see. However, they do have long, sensitive hairs on their heads and bodies that allow them to feel the walls and the ground, and they have a good sense of smell, which they use to find their way to the toilet, to their sleeping quarters, and to sniff their way to new tubers. Their skin is very loose, which enables them to move and turn easily in the narrow passageways. It's almost as if they twist their body within their skin before the skin follows their muscles and skeleton. Their lips are split at the front and meet behind their incisors, ensuring that they don't get earth and sand in their mouths as they dig.

Damaraland mole-rats rarely see the sun, so they are unable to metabolise vitamin D from sunlight as we do. Neither do they drink the calcium-rich milk of other animals like we do. We humans need vitamin D to bind calcium, in order to build up our bone mass and that of our foetuses, and to provide milk for our babies. And yet, surprisingly, mole-rats still manage to get enough calcium. They absorb more than 90 per cent of the intake they need from their diets, whereas other mammals only manage around 60 per cent. The reproductively active female in particular needs a lot of calcium, because she is pregnant or nursing almost her entire life. The others need it to renew their teeth, which are constantly worn away as they dig new tunnels.

Why are only some allowed to reproduce? Why can't they all contribute to the growth of their group, so there can be more and more of them? Damaraland mole-rats can find food underground and have few predators as long as they properly seal the holes they expel the surplus earth through. It's a relatively secure way of life. But fresh food is only available for brief periods of the year, when it has rained and the earth is soft enough to dig in. At these crucial times, many of the group's members have to be out digging – sometimes covering long distances to ensure access to food, and many others are busy storing supplies in their own easily accessible storage spaces. A single Damaraland mole-rat is unable to gather enough food by itself, but together they stand a chance. Therefore, most of the population put their own reproduction on hold as they help the reproductively active pair to raise their young, until the conditions are good enough for them to leave the safety of their community and venture out to establish one of their own. The potential benefit of venturing out is reproducing and passing on one's genes, the risk is failing to find a mate and dying alone.[125][126]

I waddle down to the doctor's surgery for yet another appointment with my GP. She can only write me a sick note for fourteen days at a time – the system forces us to optimistically suppose that the nausea will soon pass. I scheduled the visit for the afternoon – I'm less often sick then – but still I have to walk slowly, moving at snail's pace to keep my stomach from turning itself inside out. It takes me twenty minutes to walk a distance I would normally cover in five, and afterwards I have to rest in bed all day. I'm prescribed anti-nausea drugs and given a sick note for the next fortnight. The welfare state is my network – everyone has their role. I'm a birthing machine, leaving the other jobs to others. The adults at the nursery look after my three-year-old, my partner goes to work, bringing home money and food; my job just now is to make this baby, to contribute a new worker, a new taxpayer, a new individual.

Week 14

Under the thick undergrowth of the western Amazonian rainforest in South America, an animal is moving about. Roughly the size of a milk carton, it looks like a cross between a guinea pig and a squirrel, with long legs, small, pointy ears, and a thin, stumpy little tail that just sticks out of the back of its slender body. Its fur is brown with a green tinge, paler on the belly than anywhere else. The green acouchi (*Myoprocta pratti*) is almost always in motion, even when extremely pregnant, as she is now. This diminutive species of rodent lives in small social groups with a strict hierarchical system. They bite each other until they bleed, to figure out who's who in the pecking order, but they also cooperate in raising young, groom each other's coats, and signal danger by stamping on the ground. The group creates paths between bushes and trees, and every group member will help to keep them open by removing roots, stones and other objects that are in the way, because these routes help them escape large predators. They build nests, either in shallow depressions they dig in the ground, in rotten tree trunks, or in other animals' abandoned dens, and they gather nuts, which they store by burying them.[127]

Within the group, the animals form monogamous pairs; the male flirts with the female by weeing on her, chasing her about and stamping his feet until she either rejects or accepts him – in which case, the bond is formed and they are a pair. There's no point in the males making a move on a female when she's not in oestrus – it would be physically impossible for anything to come of it.[128]

The green acouchi's vagina is covered by a membrane that only opens when she is in oestrus and when she is in labour. The latter is what she's preparing for now. As her vaginal membrane is opening, she makes sure the nest is safe, and that the bedding upon which the young will soon lie is soft and dry. This membrane could be equated to the human hymen, which for a long time was used to prove whether a woman had or hadn't had intercourse. Except that it does no such thing: the hymen is not, as once believed, a membrane that covers the vagina and breaks during a woman's first experience of sexual intercourse, but a ring-shaped fold of elastic tissue just inside the vaginal entrance. Its presence or absence has no bearing on whether a woman has had sex or not.[129]

The acouchi's vaginal membrane is now open, but what is its purpose in the first place? It's a phenomenon found in a few rodents, and we don't know why, though there are several hypotheses. Perhaps the easiest of them to understand is that females have evolved it as a mechanism to control when they produce young, and with whom, and to prevent males mating with them when they don't want them to. There are many reasons not to mate with the first male who comes along. Producing eggs and carrying foetuses to term is a major investment, so it makes sense to select the best genes and to have mechanisms to stop any old male helping himself.

But the foremost research into another of the few species to have a vaginal membrane that is only occasionally open – the Gambian pouched rat (*Cricetomys gambianus*) – does not support this hypothesis. This rat, which has such a good sense of smell we train it to sniff out tuberculosis, and is so light it can be used to detect landmines without triggering an explosion, gets its name from its large cheek pouches, in which it can store food, in the manner of a hamster. Just like dogs, they can be trained with food: Gambian pouched rats love bananas, and quickly learn that sniffing out a mine, or a sample containing tuberculosis, or whatever it is we want them to find, will release a piece of banana they can snack on.

And it's largely thanks to them that scientists are ditching the "vaginal membrane as chastity belt" hypothesis and getting behind the idea that it's part of a mechanism that allows a female to control other females' reproductive activity. Both the Gambian pouched rat and the green acouchi live in social groups, and suppressing the sexual activity of others, via hormones and pheromones, means that those able to dominate get more resources and fewer competitors, both for themselves and their young. Every individual would prefer to determine for themselves when the best time is to reproduce, but if the alternative is having young who would have to compete fiercely for resources or be killed by others in the group, it's a good idea to know your place. Then it becomes a case of waiting for the right opportunity, such as when the dominant female dies.[130] [131]

Does a subordinate green acouchi long for children of her own? Does she get mad if she doesn't come into oestrus? Does she know she's a pawn in a social game, or will she simply get on with clearing the way through the undergrowth and burying food for future use, right up until she comes into oestrus and suddenly feels more interest in who wants to wee on her than in constructing paths? It's something I find hard to imagine. I don't have a vaginal membrane, I'm not suppressed by other females in the group, and I only come into close contact with wee when the three-year-old gets so involved in her games she forgets she's no longer in nappies.

Inside the safety of her nest, the dominant female is giving milk to small young with fur and open eyes. She washes them, making a purring, almost cat-like noise, and for the next six to eight weeks she will suckle them as they gradually gain independence.[132] After a while they'll head out into the world, where *they* will be dominated, waiting for their chance to be the one who gets to reproduce.

Week 15

I'm so hungry. The vomiting has let up a little – now I can eat and keep down some of my food in between the three or four episodes I still have to put up with daily. My body wants to make up for what it's lost. I think about food all the time. The first thing I do when I wake up, very cautiously so as to not awaken the nausea that's quivering down there in my belly, is eat dry, salty crackers. Sometimes it works, sometimes not. But as the morning progresses, the nausea relents a little, and I can try eating something else, even though I know I'll probably throw it up again. Vomiting has become routine: I can handle it – I want to eat anyway. The radio is babbling away in the background, but it's drowned out by a refrain running through my head: a voice saying "Eat! Eat!". First one piece of bread. Then another. Then yogurt with oats. A fried egg and baked beans. Leftovers from last night. I eat dinner-sized portions at every meal. In the first weeks, this foetus was fuelled by toast and Jarlsberg, eaten cautiously in small pieces so I wouldn't chuck it straight up again. Now it feels like I need to build up both my own body and the foetus's, and I want to eat all the food was previously incapable of thinking about. Prawns. Salted edamame beans. Wild mushrooms sautéed in butter. Soft-boiled eggs. Noodles with satay tofu. Pizza with chanterelles. Pasta with tomato sauce and lots of cheese. I can think of nothing but my next meal.

The common shrew (*Sorex araneus*)[133] has a rapid metabolism, eating 80 to 90 per cent of her body weight every day when not pregnant or nursing. But when she nurses, that goes up to 125 per cent. It helps that she weighs only a few grams, but if I were to do the same, it would require an unimaginable quantity of food.[134] She can mate again right after giving birth, and be pregnant and nursing at the same time throughout the summer. She gives birth, eats, nurses, eats, mates, eats, nurses, eats, gives birth, eats. The three dinners I eat every day add up, but if I were to consume the equivalent of what the common shrew puts away, our family finances would have long since run aground.

I eat on the bus on the way to the private midwife's centre: we're finally about to see the foetus for the first time. It's still four weeks until my appointment at the public hospital, but I want to be sure the foetus has a brain, a spine, a stomach, intestines – that it's viable and hasn't just smuggled its way past my mucous membrane defences. We're cooperating on this, my partner and I – he took the three-year-old to nursery, while I dealt with the first bouts of vomiting. Right now, I have a slice of bread in one hand and a sick bag in the other, just in case.

*

While the midwife is smearing the cold gel on my belly, somewhere far away, a mother beaver (*Castor fiber*) is bent double in a contraction. She sits propped up, with her tail sticking out between her legs in front of her, and gives birth to one kit after another. They're pushed out onto the broad tail before rolling onto the bedding she and the father beaver have assembled from soft, fresh plants. She washes and grooms them, and he helps her by eating the placenta and lifting the young up to her teats. Meanwhile, last summer's yearlings look on. They'll soon start babysitting their little siblings, keeping an eye on them while they swim their first strokes in the beaver lodge's indoor pool; they'll gather fresh branches and plants for them to eat, and accompany them on their first trips outside the nest.

The mother and father beaver have built a large lodge, which they dug out of the ground, built up with sticks and mud, and organised into several separate spaces. It takes four to six weeks to build, and it is carefully maintained and can ideally be used for many years. The entrance is under water, to prevent foxes and other predators coming in. The first room the father beaver enters when he returns from patrolling their territory is a dining space with an in-built swimming pool. This is where meals are eaten, and where all the family members who've been in the water dry their fur before entering the sleeping quarters further in and higher up, above flood level. A vent in the roof ensures a flow of fresh air. The very same day they're born, the young start splashing about in the pool. Their thick coats are so dense that they hold in air, forming a kind of life jacket; they can't dive or drown, and they can't swim out of the lodge's underwater entrance.[135]

The midwife scans my abdomen with her ultrasound probe, finding the foetus's little lodge, where it's splashing about in its very own swimming pool. It has a brain, lungs, diaphragm, arms and feet, and looks as healthy as can be. The foetus is shut in by my cervix – it has to stay underwater, it's not time for it to come out yet. It turns a little and we can see right between its legs: looks like we'll be having a boy this time. He's real, and for a few minutes the nausea and exhaustion are washed away. This is the child we've been waiting for: maybe all the vomiting was worth it.

Week 16

A spotted hyena (*Crocuta crocuta*) has just given birth. After almost sixteen weeks of pregnancy, she's made it through the greatest test of her life. Her genitals presented a challenge even as she was mating. This is because her clitoris has evolved in into a tube several centimetres long – a penis-like appendage that gives her power over her rivals but makes both mating and labour difficult.

Spotted hyenas are large predators that live in savanna regions across large parts of Africa. They look like something evolution has cobbled together from the remnants it had to hand. A large head with strong jaws and sharp teeth, outsize ears, a long neck with scruffy bristles, long forelegs and a large belly, before its body ends abruptly with slightly shorter hind legs and a floppy tail to top it all off. Its coat is spotted, its skin often scarred from battles, and its fur is usually matted with mud from a cooling bath in one of the watering holes in the territory patrolled by its group – which is called a band or a pack. But the most astonishing thing about these hyenas is that there's something hanging between the legs of both males and females.

Instead of an ordinary mammalian birth canal that makes it relatively easy to insert a more-or-less stick-shaped penis, the female spotted hyena has reconfigured her genitals into an external appendage that resembles the male genitalia. The outer vaginal labia have grown together to resemble a scrotum, and her urethra – the tube through which urine passes out of the body – runs down the pseudo-penis, as her clitoris is called. She wees from its tip, just like the male hyenas in her band. The pseudo-penis can also become erect, a trick she performs not when she is mating, but in social situations, to assert her dominance over other members of the band.

The hyenas live in matriarchies – female-dominated groups – and only female pups are allowed to stay in the band when they grow up. Females are larger than males, they control them, and they operate within a robust internal hierarchy, with social status passed down from mother to daughter.[136] And perhaps it's this hierarchy that has given rise to the female spotted hyena's anatomical oddity. There are numerous hypotheses regarding the relationship between the pseudo-penis and the group's power relations. One is that it makes sense for newborn female pups to resemble males because it stops them being killed. Hyena pups are born with sharp teeth and are happy to kill their siblings. Newborn females are at greater risk, both from their own siblings, and from other females in the band. Looking like a male, therefore, reduces their chances of an untimely death.[137] But the dangling clitoris is not only a potential life-saver – it also makes it impossible for a male to force mating on her. It's the female who decides, and she invites her selected male to conduct the act by pulling the pseudo-penis into her body, so it becomes an inverted tube into which he can stick his actual penis.

When the hyena who has just given birth sensed it was time, she stole away from the rest of the band and found her own cave where she could be alone with her pups for a few weeks.[138] Her pseudo-penis split during labour, and it still has a large wound. It won't heal into a tube again, but will stay open, covered with pink scar tissue. This happens to almost all females when they first give birth, and the scar tissue is a sure sign that she's had pups.[139] But she should be glad, in spite of the trauma, as she lies there licking her pups, because the risk of her or her young dying during her first labour was high. 60 per cent of pups born in first pregnancies are stillborn, because it takes too long to push them out through the long, narrow birth canal.[140] The young are larger than other predators' young at birth,[141] and they are born ready to fight. They have to be. They will not only need to compete with their own siblings, but also to defend themselves against mature females when they eventually return to the group cave, where the rest of the band's young live. It's a risky business. The female hyena has to return to the protection of the group, because she won't be able to find food independently, but this can be a death sentence for her young. The highest-ranking females often take it upon themselves to kill the pups of those lower in the pecking order, presumably to ensure that their own young have fewer rivals in the clan.[142]

But our hyena needn't worry about that just yet. Her pregnancy is finally over. The pups survived all the way

through the pseudo-penis, and the birth was a success in spite of – or perhaps because of – the splitting of her reconfigured clitoris.

The spotted hyena is the only species we know of that mates and gives birth through a dangling, penis-like clitoris.[143] How could it have evolved such a different structure from mine? That we don't quite know how this happened is perhaps not a huge mystery, given how little we know about clitorises in general. Female genitalia are under-studied in comparison with their male counterparts, both in humans and in other species,[144] and the question of how the spotted hyena's clitoris grew into a pseudo-penis is no more of a mystery than how clitorises in general look and work.

The human clitoris has passed in and out of medical books on genitalia. Throughout history it has often been seen as too insignificant to warrant inclusion. And for a long time we believed it to comprise only the small knobble located above the vagina: a little pea,[145] not as large and swollen as the penis. That is incorrect. The human clitoris is ten centimetres long and stretches up into the body, and the pea is only the head of a four-legged organ made of a spongy tissue called corpus cavernosum, which surrounds both sides of the vaginal opening and can swell up (just like a penis). It wasn't until 1998 that the first anatomical study of the human clitoris was published, and in 2005 the world was finally able to witness its full anatomical glory through the work of Australian urologist Helen O'Connell.[146]

Until very recently, we believed that the clitoris had eight thousand nerve endings. That's an impressive number, and we have used it to show how sensitive evolution has made this little organ, but it also shows how little we actually knew. The number eight thousand comes from cows, not humans. It wasn't until 2022 that someone counted the nerve endings in a human clitoris, reaching an average figure of 10,280.[147]

*

I won't be giving birth through my clitoris, but it does surround my birth canal, and its role during labour is something we know even less about. One thing we do know is that it receives internal stimulation both when something penetrates the vagina and when something tries to push itself out through it. As a baby's head is pushed down towards the birth canal, it stimulates the clitoris at various points during the process. This may well be connected to a series of mechanisms during pregnancy. The so-called Ferguson reflex – when cascades of the hormone oxytocin are released by the pressure of the baby's head through the birth canal, leading to the uterus contracting and continuing to press the baby out – is an important aspect of labour. It means that the birth is in progress. Some believe that this reflex is initiated not by pressure on the cervix, as is widely held to be the case, but on the internal parts of the clitoris. In addition, the clitoris may play a part in further widening the pelvis, by being connected to certain muscles that are in turn connected to the coccyx, which is pushed outwards to enlarge the space through which the baby will be born. And perhaps the four legs of the swollen clitoris also protect the back of the baby's head on the way out.[148] My clitoris played its part when the sperm cells came in – perhaps it will have just as important a role when the product of the egg and sperm cells is ready to come out, nine months later, almost twenty-five weeks from now.

Week 17

Geoffroy's tailless bat (*Anoura geoffroyi*) is found across large parts of Central and South America, where it flies around at night and hangs upside down in its hiding places during the day. It's a medium-sized bat, weighing up to fifteen grams, and has brown fur, a mild underbite and a funny, thin little triangular nose that points vertically upwards. Its long, narrow tongue can stretch far out of its mouth, making it easy for it to access pollen and nectar from blossoms and catch insects in the air.[149]

For such small creatures, bats have very long pregnancies, much longer than most other small mammals. The average weight of an adult bat, if you count every species, is twenty grams, but they are pregnant for three or four times longer than other mammals of a similar size. They are also the only mammal that can really fly, not just glide, and major evolutionary changes have taken place along the route from front legs to wings. Perhaps the bat evolved to be smaller so it could carry its own weight on its wings, while retaining the reproductive pattern of a larger animal, with a long pregnancy and a litter of just one.[150]

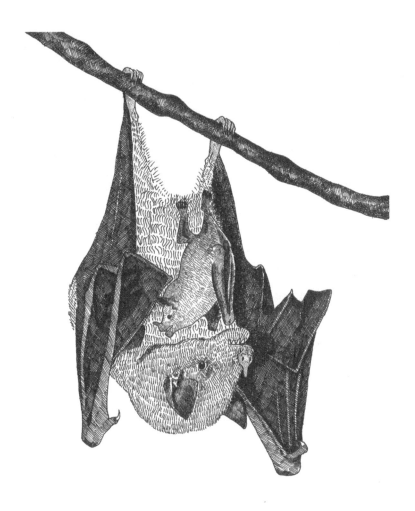

A bat's fingers are trapped inside the thin skin of her wings. Only the thumbs stick out beyond them, each with a single sharp claw with which to hold on to things. Between her fingers stretches the patagium, the skin-like membrane that forms her wings, catching hold of the air and pushing her forward with every swipe. By contrast, my fingers are distinct and separate, built for precise work. Originally this might have meant picking up grains from the ground or making arrowheads, but I tend to use them for hitting the keys on my laptop. These days I find myself cupping them, so I have something to vomit into if I don't make it to the toilet in time, but since there's no skin between them they're not watertight and it drips as I walk.

When the Geoffroy's tailless bat is not flying, she lives her life upside down, hanging from her short hind legs. When giving birth, she probably turns round to hang by her thumbs, using gravity's help to push out the pup, bending her knees to catch it with her legs and their strong claws. After seventeen weeks with a foetus growing in her uterus, she will finally be free of its enormous weight, which can reach up to 45 per cent of her own. That's like my baby weighing almost thirty kilos when I give birth – how is she able to fly? When the pup is finally out, it will use its little hind legs to hang on to her, and she'll wrap it up within the safety of her wings, where it will latch on to a teat and grow even larger outside her body. So how does she manage to give birth to such a large pup? How is she able to push it out?

In the majority of bat species, the male's pelvis is a fused-shut ring, while the female's is open. The two front portions of the pelvic bone don't meet as ours do, being connected by a ligament instead, a band of connective tissue. This makes the circumference of the pelvis much, much larger, so the enormous pup can slide out through an opening which in human females is much too narrow.[151] Bats hang by their hind legs when they sleep, and while awake they live their lives horizontally, facing downwards, which means they don't need a narrow pelvis that can support them on their hind legs and prevent their intestines falling out as they stand upright.

The mother bat hides her pup with her wings. It will grow quickly now, despite its already large size. Pups are born with teeth, and strong thumbs and toes to cling on with. They're vulnerable until their wings can bear them, and they have to gain strength quickly. But it's not just the need to fly on their own after a short while that leads to bat young being born so large. The bigger the body, the easier it is to regulate body temperature, and the larger the pup, the lower the risk of it freezing to death.[152]

Week 18

It's morning in the Peruvian rainforest. Sunlight filters down through the thick canopy, the air is warm and humid. A broad spectrum of organisms live among the branches and lianas, from bacteria, fungi and mosses, to insects, birds and mammals. A little reddish-brown primate is sitting on a branch. Its body is less than half a metre in length, but the long tail makes it look much bigger. A San Martin titi monkey (*Plecturocebus oenanthe*) is about to give birth. Soon her work will be done, and the father will take over the primary caring responsibilities. This little primate, which is only found in a few locations in Peru, weighs about a kilo. She lives in a monogamous, often life-long relationship with a male. Their group consists of themselves and a few young.[153] Right now, her small brown body is hunched over. She's found a thick, horizontal liana that's good to sit on, with many small branches to hold on to. She's stopped eating, because her labour is now under way. The long tail hangs down to balance the little body. The male grooms her fur and checks her vaginal opening several times during the labour. Is there a baby coming yet?

One hour and many contractions later, the head can just about be seen peeking out between her legs, and within ten minutes the whole body has emerged, without any help from her hands. The baby clings to her thigh, and she starts to eat the placenta, which was delivered immediately after the infant. The male is on hand – he's been close by her all this time, grooming her fur and taking care of her – and now he begins to lick the newborn. The other two young in the group, a nine-month-old female and an eighteen-month-old male, study the new family member from a distance. They must be curious about this new arrival, who is set to take their place in getting the most care and attention, pushing them further out of the innermost circle.

After a few hours, the newborn lies at the female's breast to drink milk, and later in the afternoon it is taken to the group's regular sleeping quarters in a thicket of bamboo. But, starting the next day, the father will be the one who carries the infant, and it will be with its mother only when it nurses, just a few times a day. It also remains with the mother at night, tucked up close to her body, but only until it is two months old. Then the father will take over the night shift too.[154] After two and a half months, the infant will be largely capable of moving about, finding food and eating on its own, but the male will still carry it over longer distances until it is five months old.

Of all monkeys, the male titi takes on the most responsibility in caring for his young. He carries them, grooms them, makes sure they don't fall out of the trees in which they live their lives, gazes deep into their eyes, plays with them and ensures their big siblings don't get too boisterous. He can't nurse them, so he still needs the mother for that, but he has the infant 90 per cent of the time after it's born. Why is he the one who provides most of the care? It could simply be that the female needs to dedicate time before and after the birth to catching insects, which provide more sustenance than the fruit and plants that make up the majority of the titi's diet. With the male taking over the lion's share of the caring responsibilities, she can spend more time seeking out the most nutritious food, making it possible for her to produce milk and build up resources for her next pregnancy.[155] A birthing interval of only nine months between each infant suggests that the division of labour works well for this species. Studies on titi monkeys in captivity show that the young become more stressed if the father is briefly absent than if the mother temporarily leaves. He is the primary caregiver, even though he can't give the infant milk.[156]

Does the mother feel sad because her baby would rather be with Dad, or does she think it's nice to be free, to have some time to herself? Is he proud to be the animal kingdom's foremost proponent of gender equality, or does he feel like his masculinity is threatened? In my den, there have gradually been fewer and fewer calls of "Mummyyyyyy!" from the three-year-old, and more and more appeals to Dad. I've consoled and carried, nursed and cradled, and even if we've tried to divide everything equally, I've always been the first port of call when a grazed knee needs kissing better, or colouring pencils are required. But now I'm not as available: my brain is somewhere else, my embrace is full of a growing belly, and her other primary caregiver has risen in status.

Week 19

If I drink a glass of juice and lie completely still on the sofa, I can just make out the odd kick. It feels like gas or butterflies in my tummy, but I soon begin to recognise it as the movements of the foetus: a flailing arm, a kick or a flip in a strange place deep inside my abdomen. It's not like a touch to the skin, when it's easy to know exactly where on my body someone has brushed against me. The foetus's movements feel like something sloshing about in an unspecified place inside my growing bump, like waves in a viscous mass moving outwards from the centre.

On the sludgy edge of a large lake somewhere in South America sits a thing that looks a little like a half-inflated whoopee cushion. It has a large, beak-shaped mouth that stretches between two rigid-looking arms in front and two meaty legs sticking out at the back. It's a frog[157] that goes by a deceptive name, the Surinam toad (*Pipa pipa*), and there's something kicking her from the inside, but it's not from her belly that this kicking comes.

Out of her back, tiny froggy arms are emerging, along with froggy legs, and here and there a little head. She's giving birth to babies – each one is gradually forcing its way out of its hole, where it's been taking shelter ever since the female frog and the male frog mated. The mating took the form of an underwater dance involving cartwheels and somersaults, as he clung to her back and made sure all the eggs she'd pushed out of her cloaca were stuck fast, while simultaneously bathing them in sperm. The skin on her back was already swollen and primed to receive the eggs, which come equipped with their own glue to ensure they don't fall off. Just as my egg attached itself to the lining of my uterus, the Surinam toad's eggs stick to her back, and over the course of several days, her skin grows up around them to completely enclose them, each in their own little chamber. They've been there a long time now,[158] first as eggs, then as wriggling tadpoles, still each in its chamber, with a little hole facing out into the world, before undergoing a metamorphosis and becoming small frogs, ready to emerge, swim up to the surface and fill their lungs with air for the first time.

Usually, female frogs lay their eggs in a pond while the male clings to her and squirts sperm over them. After that the young have to fend for themselves – first as eggs, then as tadpoles, and finally as miniature frogs that will one day grow large enough to reproduce in turn. The Surinam toad and the group of frogs they belong to do not do this. They live in areas where water can dry up – the ponds are often small, and the eggs and tadpoles might dry up in turn and die. Or they might be eaten by fish, insects or other frogs. It's safer for the adult females to take their brood with them, caring for them as I'm caring for

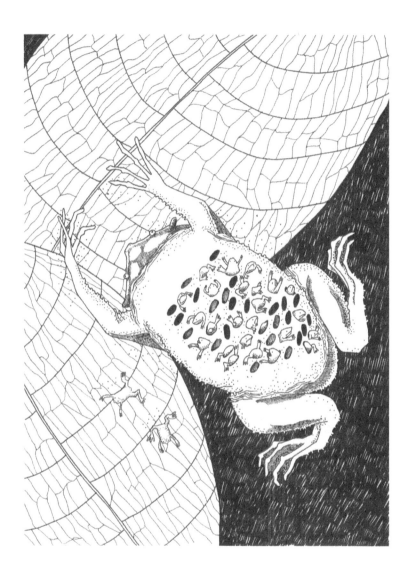

my baby, inside their body. Though we're far removed from an evolutionary perspective, each on a different branch of the tree of life, we've both found a way to care for our growing embryos by enclosing them in our own bodies.[159]

*

As far as we know, the young of the Surinam toad don't receive nourishment from their mother, and they certainly don't have an umbilical cord like my foetus does. Instead, they get oxygen from the water, since the mother is not able to give it to them through her skin. But even though there is no evidence she provides sustenance in the way the male seahorse does through the brood pouch on his back, or the male Southern Darwin's frog does in its vocal pouch, something clearly does happen inside the egg's chamber. Eggs that don't attach to their mother's back, and instead end up floating in the water, die after a couple of days. The chambers do not have a placenta, and they are not a uterus in usual sense, but an interaction has evolved between the egg and the mother that allows the eggs to survive within the skin on her back, but not in open water. We just don't know the details yet.[160] Studies that also examined the skin of a related species that births its young as tadpoles and not as miniature frogs have shown that some of the same hormones are involved when the Surinam toad's eggs sink into the mother's back and are enveloped by it as when my egg embeds itself in the endometrium.[161] It appears to be the same evolutionary building blocks making it possible for me and the Surinam toad to carry our young, even if the way we do it is different.

That's not as surprising as it might sound. It's common for similar characteristics to evolve in completely different species if they bring about the same evolutionary advantages. The wings of a bat and a bird have a very different basic structure, but the function is the same – to fly, to have the opportunity to exploit the air and conquer a new element. The Surinam toad and I both want to protect our embryos, and we carry them with us instead of laying them as eggs and leaving them. But we're still different – she has many young at once, and cares for them only until they are tiny frogs. She's done now, after nineteen weeks, while I'll be caring for my child for nineteen years and beyond.

Week 20

The eggs of our Namaqua chameleon are hatching. The tiny newborn reptiles climb straight up into small bushes for safety. They need to grow in size and strength before they can defend their own territories, so they're permitted to stay close to their parents for a while. But there's a catch: it doesn't often happen, but they're sometimes eaten by their mother – perhaps it's not always easy to tell the difference between your own children and a tasty snack. While the eggs have been incubating in the hole she dug for them in week five, the mother chameleon has been free to eat, to move without the burden of their extra weight, and to lay further clutches. It's possible she watches over the eggs, as suggested by the fact that she sometimes lays them in an extension of her own burrow, and we know she does a daily patrol of her territory – perhaps to ensure that no predators come and dig them up.[162]

The Namaqua chameleon is dependent on the sun to warm her body. She's cold-blooded, and cannot produce her own heat, so as the sun comes up, she creeps slowly out of her burrow, turns the side of her body with the largest surface area to the warming rays, and makes her skin black so as to absorb the most heat. When the sun is at its highest and it becomes too hot, she makes her skin white to reflect its rays. Right now, I feel like I have a radiator inside me – the foetus is like a little heater. It's the hormones and the increased blood supply to my skin that's doing it. This winter I don't need to wrap myself in lots of layers – I walk with my jacket unzipped, letting my body heat radiate up towards the sun.

Week 21

The wild sheep (*Ovis aries*) that roam the coast of Norway are an ancient breed – domestic animals that live their whole lives out in the open, grazing on heather moorland and largely fending for themselves. They are owned by farmers who keep an eye on them, moving them between grazing areas and ensuring they have dry places where they can shelter from the elements, giving them extra feed in the winter if needed, and making sure lambing goes according to plan. Now it is spring, and one female, among many others, is preparing to usher in the next generation. Through the long, dark winter, a lamb foetus has been growing in her uterus. She has been using the fat she put on through the autumn to build the lamb, gram by gram. She weighs approximately the same just before lambing as she did when she was nice and fat from the fresh grass she eats in summer, but the kilos have shifted from her to the foetus.

The hay, seaweed and heather she relies on in winter are sustenance feed, not fattening-up feed, and now her heavy body is finally about to get some fresh grass and to free itself of its burden – soon she'll be herself again. The sheep flock together, even when lambing season is approaching. They move slowly for the moment, but as soon as the young are born, their bodies are suddenly lighter; both the ewe and her lamb are built to be able to run from danger – they need to be quick on their feet to survive, right from the moment of birth. Soon they start to drift apart – they don't stand in dense groups while they're lambing. A tame animal with wild instincts, our ewe moves away from the flock and finds a calm, safe place to give birth, a place where she can get to know her new lamb without being disturbed by eager aunties or hungry eagles.

In the hours before the lamb is born, the mother alternates between standing and lying down. She has strong contractions – her uterus is working to get the lamb out – and over the course of up to six hours, the birth canal opens fully and the muzzle and front hooves of her young begin to emerge. Ideally, she will be standing when the final contractions push the lamb out, so it can tumble onto the ground and she can turn round quickly and ensure the foetal membranes aren't covering its airways by licking its nose clean.[163]

All animals have hormones racing through their bodies, chemical substances that help regulate how we function. The oestrogens that were so essential when it was time for my ovary to release an egg into my uterus are steroid hormones that regulate fertility, among other things, both in sheep and in humans. And they got their name from the way sheep react to their sinuses being full of parasitic larvae.

The sheep bot fly (*Oestrus ovis*), as the name suggests, uses sheep as hosts for its larvae. After mating, the female fly carries the eggs inside her body until the larval stage. Then she locates a sheep and quickly deposits the larvae in its nose. It happens so quickly that before the sheep has had a chance to react, the tiny millimetre-long larvae have crept into its nostrils and up into its sinuses. Once safely stowed away, they gobble up mucus from the sinus walls and grow to a length of two centimetres. And when they're mature, they crawl out of the sheep's nose, fall to the ground and pupate.[164] Sheep get anxious when they hear the buzzing of these flies, running, kicking, and often eating less – they don't want the larvae up in their nose, inside their bodies. For farmers in times gone by this behaviour was virtually indistinguishable from the how the animals acted when they were on heat, that is, when they wanted to mate. Another word for being on heat is *oestrus* – the same word as appears in the fly's name. When the hormone that brought sheep into oestrus was discovered, naming it was simple. It was called *oestrogen*, after the little fly that makes sheep act as though they're on heat.[165]

At about this time, in my body, oestrogen production – which was previously handled by the corpus luteum – is taken over by the placenta. Working together with the foetus, it sends out hormones that cause the uterus and the tissue of the mammary glands to grow,[166] ensuring that I am able to sustain this new life.

Week 22

Our giant Pacific octopus, who we first met in Week 8, now flushes water over her eggs for the last time. Once a large, muscular animal with red-orange skin, a fierce, intelligent hunter, she's been reduced to a shadow of her former self. Her skin is grey and covered with large lesions, her muscle mass has shrunk in line with her decreasing food intake. Over the last few weeks, the small spawn in her thousands of eggs have been moving more and more. The baby octopuses have been spinning around increasingly quickly inside the soft egg membranes, and now holes are forming in the balloons, and the two-centimetre-long, fully developed, bright-red young stream out into their mother's cave. With her last ounce of strength, the mother octopus flushes them out of her safe hiding place, up into the open water, where they will find food to eat, or be eaten. The vast majority of them will face the latter fate, with only a few surviving the three years it takes them to reach sexual maturity themselves; the rest will become fish food, shrimp food . . . or octopus food.

Back in the hiding place, the long strands of empty eggs

float in the water above a grey corpse: a mother who has given everything for her young and is no more. A crab approaches this former crab-devourer, and, sensing that she no longer poses any threat, as the carrion-feeder it is, starts to eat the thing that would once have eaten it.

To reproduce only once – to die after just one brood – is that greater, nobler or more important than what I'm doing? I'm dragging myself through a second pregnancy, at the same time as looking after a three-year-old who's about to turn four, who needs all my love and attention, who will need me for another twenty years, need me to be a grandmother if she has children of her own. Octopuses are semelparous, I'm iteroparous – these are terms that describe two different reproductive strategies. Either putting all your eggs into the basket (so to speak) of a single reproductive cycle, after which you have nothing to live for, or reproducing little by little, having a brood here and a brood there, and perhaps looking after them for a shorter or longer period after they've hatched or been born.

The strange thing is, if you were to remove a little gland located right behind the octopus's eye – the optical gland, equivalent to our pituitary gland, a little hormone-producing hub in our brains that's essential to life – the octopus doesn't die after laying its eggs. It will survive and can even mate again. But it's not allowed to – its death is predetermined by its hormones.[167]

Semelparity – explosive reproduction, followed by death – has evolved several times in species and genera that were previously iteroparous, that had offspring several times before they died.[168] Most salmon (*Salmo* spp.) are like the octopus – many plants and insects also, and there's even a little genus of marsupials in Australia, *Antechinus* spp., that only have young once.[169] Dying can have its advantages. It could be that the octopus and the salmon are able to channel all their resources into producing vast quantities of spawn because they don't need to save anything for when they themselves grow old. They can reproduce at the best possible moment and go all in; they don't break off halfway, like the common eider occasionally does, or expend energy on protecting themselves against parasitic foetuses, like we humans have to. Evolution has caused species that survive through the generations to do so in different ways, depending on the genetic material that exists in the population, and the environmental conditions under which they live. For some species, a single round of reproduction is preferable; with others the young have to be cared for and watched over if they are to survive: they have to be spread out over time, they have to be supervised, they need to learn to find food for themselves, protect themselves from danger, find the way to school, figure out how online banking works . . .

Evolution has also ensured that there is variation in the mechanisms that result in semelparous species dying after reproducing. In the case of salmon, we believe that the cost of producing sex cells and finding a partner are so high that death is unavoidable,[170] but that's not true for octopuses.[171]

There are probably other evolutionary mechanisms that have led to octopuses dying after their eggs hatch, but we haven't identified them yet. Perhaps it stems from the fact that these creatures are fierce cannibals – those cute little red-orange octopus babies with big eyes, floating around in the water practising their arm coordination will gladly eat each other, given half a chance. And dying when your young have just hatched is a pretty effective method of ensuring that you won't eat your own babies, especially given that octopuses never stop growing: and if they never died, there would be no space for new, young octopuses, who wouldn't be able to grow as large. Perhaps that's why the giant Pacific octopus mother is no longer even a memory, her remains devoured by hungry fish and crabs.[172]

As she gradually fades away, I come back to life. My body is growing, and though the foetus is still in control, I'm slowly emerging from the dark hole I was in, as the nausea loosens its grip a little. I'm taking three different anti-nausea drugs, four times a day, and it means I'm able to think, to gradually start using my body for more than just sleeping, eating and vomiting. The fog lifts, I'm capable of relating to the world around me, of thinking how Christmas is just a couple of days away, that the year has turned a corner and the days are getting lighter. I'm over halfway now, and it feels like I might just make it.

Week 23

I'm still queasy and exhausted, but when I'm done vomiting, I can summon up precisely enough energy to collect my daughter from nursery. My body feels useless, the smallest thing can send me over the edge, making me throw up – the anti-nausea drugs only help to a certain extent, and only on some days am I able to manage a little work. I take twice as long as usual to walk to the nursery, because I have to keep stopping along the way, even though it's only a couple of hundred metres door-to-door. The light outside is sharp, and I've been inside all day. I meet another mum on the street and she asks how it's going. She says she had so much fun the last time she was pregnant, but I just feel imprisoned by my own foetus. Not that I can say that out loud.

In the lobby, where the children hang up their coats, it stinks of filthy rubber boots, sweaty underclothes and foetid socks. There are too many smells, so I breathe through my mouth to keep the nausea at bay. The room is right next to the toilet, and every single day I find myself hoping no-one needs to go while I'm doing pick-up. My three-year-old throws herself on the floor and refuses to put on her snowsuit and it takes just that bit too long – my nose is caught in a prison of smells and my body revolts – it's protecting my foetus. I vomit into the tiny child's toilet and hear one of my child's friends shout "Oh!" at the strange sounds. When I come out again, one of the staff has got my little girl dressed and is looking at me with sympathy. I'm longing for my bed, for my dark cave.

We dawdle on the walk home. I may be slow, but the three-year-old is even slower: she wants to collect sticks and look at pretty stones. I want to waddle home, collapse onto the sofa and wait for her dad to come and save me from my own child, so I can be nauseous in peace.

The Thomson's gazelle (*Eudorcas thomsonii*) gives birth at twenty-three weeks.[173] It's one of the world's fastest animals, reaching speeds of up to ninety kilometres an hour, only a little slower than the cheetah, its primary enemy. The slender female, slightly taller than a golden retriever but only half as heavy, appears visibly pregnant, but not in a human, panting, swollen-faced way. Her belly is a little rounder, she's a little heavier in the body when running from predators, but her legs are still long and willowy, and her ears, with their internal black-and-white patterning, are on high alert. She resembles a ballerina, even as she withdraws to give birth. Moving a short distance away from the herd, away from the open habitats of short grass that they prefer, she finds an area with slightly taller vegetation where it's easier for the fawn to hide.

She gives birth standing up, and the fawn falls to the ground with a thump. In contrast to many other hoofed animals, the young gazelle cannot run right away. The mother's strategy is to hide the baby fawn in the long grass, where it lies completely still and waits until she returns to nurse it. Where the young of some species are followers – they get up straight away and follow the mother and the flock – Thomson's gazelles are known as hiders. The young remain concealed in the long grass, while the mother stays close enough to keep watch, but not so close that predators will be able to detect the fawn's whereabouts. Infant mortality is high in many gazelle species: 50 per cent of all fawns die, mostly taken by predators. The fawn will gradually want to move about more, and the mother will take it with her into the common protection offered by the herd, but right now it's best for it to lie completely still, crossing its hooves that a predator won't catch its scent and burst into its hiding place.[174]

I'm not remotely graceful right now, the last thing I would compare myself to is a sprightly, nimble gazelle, but I can comfort myself with the knowledge that infant mortality in humans is nowhere near 50 per cent. I don't need to be able to run from a predator, whether pregnant or otherwise. And while I'm currently on the slow side, I'll soon get my body back and be master of my own life again. I'll be an efficient mother-of-two with a job, accomplishing everything from nursery drop-offs to cooking dinner. I might even be able to enjoy my own hobbies again, once the baby is born.[175]

Week 24

Over an area twice the size of Paris, in the Weddell Sea near Antarctica, a vast blanket of nests are tightly packed on the seabed.[176] They belong to a colony of Jonah's icefish (*Neopagetopsis ionah*) that has made its home here. A male swims low over his nest, guarding it as others guard each of the 60 million or so nests in the area. There are 1,700 eggs beneath him, laid among small stones and gravel, in a depression in the seabed that he has dug with his elongated, spade-like lower jaw. He makes sure no-one eats them, clears away debris and parasites, and carefully monitors their progress.[177]

This remarkable breeding site was discovered by chance in 2021. It's the world's largest colony of fish eggs, belonging to a species we know little about, in an environment – the ocean – that is so close to us, yet so hard for us to understand. We can't breathe underwater, and we wouldn't survive the pressure of the deep sea. More than 80 per cent of the world's oceans have yet to be explored – we know more about the moon's surface than the deep-sea floor of our own planet.[178] Even now, in the 21st century, it's possible for a research expedition to discover an enormous city under the water – a society of fish dads watching over their eggs.

Around our male Jonah's icefish, we find nest after nest, and male after male caring for his eggs. Many of them have died in the process – the long incubation phase of three to six months[179] wears them out. Their remains lie slowly rotting on the seabed, unless they've already been eaten by the seals and other organisms they share their living space with.

I'm at home, sitting in my nest: an apartment in Oslo surrounded by many similar apartments. There are over six hundred thousand of us in my city, but fewer than ten thousand babies are born each year. These nests are not just for eggs – and it feels like the nearest parent-to-be is far, far away.

Week 25

Seventy-five million years ago, in an area covering what is now Montana in the USA and Alberta in Canada, a clutch of duckbill dinosaurs are in the process of emerging from their eggs.[180] [181] Among the bracken in a mixed woodland of evergreen and deciduous trees, a new generation of dinosaurs from the species *Hypacrosaurus stebingeri* fight their way out of the nest mound. The eggs have been kept warm by microorganisms in the soil that create heat as they break down dead plants and animals, just as in the nest of the Australian brush-turkey.[182] They weigh a little over four kilos at the time of hatching,[183] and, once freed, the baby dinosaurs will grow from around 1.5 metres long[184] to full length of about nine metres, with a lifespan of around thirteen years.[185]

H. stebingeri was a plant-eater, with a long neck, strong back legs and smaller front legs. On the top of its head, it had a hollow, bony crest. Perhaps this was used to make sounds, or as an identifying marker. Its mouth was like a duck's bill – hence the species' vernacular name. But it wasn't a beak: inside its mouth it had a full array of teeth that could crush the fibres and cells of plants.

Were the little dinosaur babies met by a parent when they came out of the nest? Did their mother or father make sure they weren't eaten by predators, that they knew what to eat and how they should live their dinosaur lives? Which of the parents made the nest, and did they share any caring responsibilities, or did they abandon it and use their energy to lay more eggs, producing many young that had to fend for themselves, rather than a few that needed caring for?

The truth is we don't know, but the evidence suggests that another member of the duckbill dinosaur group, *Maiasaura peeblesorum*, appears to have protected its young. The *Maiasaura* lived a little before *H. stebingeri*, and got their name because people imagined them to have been good mothers – *Maia* comes from Greek and literally means "good mother". These dinosaurs were named when fossils of adults were found together with eggs and newborns. The eggs of *M. peeblesorum* were smaller than *H. stebingeri* when the young hatched, weighing in at around a kilo,[186] and the incubation period appears to have been shorter. Nests with many *Maiasaura* young have been found, encompassing eggs, newly hatched young and larger juveniles that had been alive a while, their teeth worn down from their diet of bushes and leaves. They were probably cared for. Why else would they keep returning to the nest, if there were no adults there to protect them from predators?[187]

Was the nest a prehistoric form of daycare? Did dinosaur

parents leave their young there because they had to go out to find food, to earn their crust, just like we do? Did the children develop their own sense of humour, their own culture, there in the nest, like my three-year-old, who comes home and tells me a joke about a poo crossing the road? Did some of the little dinosaurs prefer to be with certain of their peers more than others, did they have best friends, did the bigger young in the nest have lessons where they practised skills for the next stage in life, skills they would need if they were to survive long enough to reproduce themselves?

Is the *Maiasaura* mother waiting by the nest? Is she protecting it as the Nile crocodile, her distant, distant relative, protects hers seventy-five million years later? Dinosaurs cared for their offspring and their eggs in different ways – just like birds, their descendants, do today. They were around for such a vast stretch of time, and were such an enormously varied group, that it's impossible to generalise according to finds relating to one or a small number of species. Some dinosaurs' eggs were hard and robust, and others soft, with clay-like shells. Nests have been found full of eggs with no adult dinosaur nearby, just as fossilised specimens from several species have been found in close proximity to their nests, even lying right on top of them, like some birds sit on theirs today. Perhaps one day we'll discover cuckoo dinosaurs who laid eggs in the nests of other species, just as the cuckoo and the cuckoo catfish do now.[188]

Week 26

A female brown-throated sloth (*Bradypus variegatus*) is about to give birth, and it will take place, like almost everything in her life, while she is hanging in a tree. She only comes down to the ground to defecate, and even though it may in some ways feel similar, that's not what she's about to do now. After half a year, she's finally done carrying the foetus inside her. She hides herself deep in the trees, somewhere in Central or South America, her greenish-brown fur hard to detect among the tall trunks, with only her little face sticking out, the two panda-like black patches around her eyes stark against the surrounding pale fur.

It's not easy to tell when a sloth is pregnant – her belly doesn't get particularly large. But from the moment the baby is born, it spends months using her as a hammock, and that's pretty clear to see, making it easy for scientists to determine her reproductive status. She hangs horizontally from her long, curved claws while giving birth, and the baby lands on her stomach: perhaps she guides it into place with one arm. We would never be able to hold on so long, but the sloth's body is built for this kind of thing. Its finger bones grow out beyond the skin of the fingers along with the nails, together forming large claws that make it easy for it to hang from its arms and legs without using any additional energy. The little baby has open ears and eyes, and nurses for only four weeks, even though it will be carried by its mother for the next six months, lying stomach to stomach, using her as a mattress.[189] [190] [191]

The brown-throated sloth's diet consists of the leaves of only a few kinds of trees, leaves that give so little nourishment and take so long to digest that her whole life has to be lived slowly. She is dependent on the sun to warm her slow body, which, with its low temperature, puts her in the same boat as a cold-blooded reptile. Her small size – she grows no heavier than 6.3 kilos – helps her to shelter from predators out on the thinner branches of the canopy. She could stay out there all her life, but she doesn't. Once a week she climbs slowly down to the ground, down to her predators' and enemies' level, to defecate. Why does she expose herself to danger in this way? You might think she would prefer to climb down to the ground to give birth, instead of risking losing her baby by birthing it hanging from a liana,[192] but instead, it's her faeces she takes great care to leave down there. And it's all because of what she has living on her body. The sloth's fur has a green tinge that comes from the algae that grow in it. These algae give her both protection and extra nutrition, offering sustenance she doesn't get from the leaves that otherwise make up her diet.[193]

At the same time, there's something else close at hand that wants to reproduce, because her fur offers a home to a whole society of moths (Microlepidoptera), who also happen to need sloth faeces in order to reproduce. While she's down on the ground, ideally in her favourite toileting spot, she leaves behind the remains of the previous week's digested leaves. That's when the moths fly up out of her fur and down onto the droppings, where they lay their eggs. The eggs turn into larvae, which eat the droppings before undergoing metamorphosis and becoming new moths, which take to the air and find themselves a sloth whose fur they can live in.

The moths live and mate and die in the fur, and provide nutrients for the algae that the sloth in turn depends on for vital sustenance. The sloth needs the algae, the algae need the moths, and the moths need the sloth to defecate in a way that enables them to fly onto the poo, lay their eggs and fly back to the sloth's fur.[194] Perhaps that's why she relieves herself on the ground, but gives birth in safety up in the trees. It's not a question we can answer, but in any case, now she has a little sloth baby on her tummy, one that will need milk and care and algae and its own moth colony before it's ready to go its own way, in time for the mother sloth to have another baby in a year's time.

Week 27

Right by one of the world's strongest whirlpools, the Saltstraum maelstrom near Bodø in Northern Norway, a male Atlantic wolffish (*Anarhichas lupus*) lies curled around an egg ball, a compact mass of several thousand eggs. The vortex here is formed by tides pulling a huge quantity of water through a narrow strait, and with it the oxygen and nutrients to support a large food web of different species. High up in this web is the huge Atlantic wolffish, with its sharp teeth and strong jaws. He looks angry all the time, the corners of his mouth drooping and his teeth sticking out. Right now, his grey-blue skin is looking more washed-out than usual, and if a fish is capable of looking tired, that's how he looks.

And no wonder – he's been lying still for a long time now, and the tiny young in the eggs are starting to move about more and more. It's been getting increasingly cramped for them in there, and soon the egg membranes will split and they'll pop out and make their way into the world alone. For the last few years their father has been excavating gravel and mud to dig out a hole under a large stone in a suitable location. And in recent months he's been rewarded for the grand job he's done in building a safe home, with the opportunity to sit on his eggs in peace, free from the threat of being eaten by others.

It's easy now to see the eyes through the egg membranes, and the embryos already have teeth. They also have their yolk sac, which they can live on for a while before they have to find their own food. Meanwhile, the father has lain curled around his egg ball for a long time, and his energy is fading – he hasn't eaten since his partner laid the eggs seven months ago.[195] He's been there throughout the long winter, with no food, motionless but for the occasional change of position, ensuring that the eggs aren't eaten by crabs or other fish passing by. For months prior to this he lay in his hole alongside the female, after ensnaring her as she came up from the deep sea in the spring, just a little later than he did.

The hole is the fruit of his labour, and perhaps she chooses him specifically for his building skills. It might also be he who makes the choice, after several females have courted him while he lies inside, observing the show they put on. We don't know, but either way, after a few days of intense flirting, the wolffish couple take their time, spending many months in each other's company before she's ready to lay her eggs. They curl up together in the hole, venturing out to hunt sea urchins and starfish and bringing the food home like a kind of romantic take-away dinner. When the time comes, he transfers his sperm to her by some means we know little about, and the eggs are fertilised internally before she lays them in a large ball, all stuck together. Then she hands responsibility for them over to him. She's done her job now, used up all her energy, and like the emperor penguin, she leaves the eggs in the male's care, moving off into slightly deeper water. Here she can find food and build up her reserves again through the long winter, before returning to another hole, or perhaps the same again, mating with the same male: multi-year Atlantic wolffish pairings have been observed in the Saltstraumen maelstrom, but we know little more about these animals' habits. Most wolffish research has focused on the possibility of preparing them for human consumption, encouraging them to mate in captivity and getting their egg balls to hatch without a male being present. We don't dedicate much time to what makes a female choose a particular male, how they divide the labour of raising their young, or how exhausting it is to be responsible for the next generation.

Can the male wolffish have a good moan to his mates in the nearby holes about how tiring his duties are? Does the female tell him before she leaves that it's high time he stopped whining and pulled himself together? Do they discuss the most tiring aspects of laying the eggs and looking after them when they meet again the following year?

When you think about it, I'm like the male wolffish. I have to stay with my young, limiting my movements while my partner is able to live his life. Granted, my partner doesn't leave – he takes care of me, brings me food – but he still has the energy to go to the cinema, meet friends for a beer, give presentations at work, take the three-year-old to the park. He lives his life, while mine passes me by. My pregnancy is a swamp I'm stuck in, unable to pull myself out. The clock is ticking, but for me time stands still.

I have more in common with the male Atlantic wolffish and the male seahorse than the females of their species. Are they exceptional super-fathers, or are we viewing them too much through the filter of human gender roles, based on which sex cells they produce? When I think of a male and a female wolffish, I automatically imagine him looking after her and being the more aggressive of the two should an intruder come along, while she is calmer, more careful, more concerned with protecting her body and her eggs. But, of course, I know nothing about how they live their lives. I can barely imagine what it's like to have gills instead of lungs, to swim up and down several hundred metres of sea, to produce thousands of babies at once and never see them again after they've hatched. Is it natural for all males, regardless of species, to have one pre-determined role, and all females another? We've evolved to be so vastly dissimilar. Just think of the differences between a fish and a human. It's strange to accept these divergences as natural, while at the same time insisting that males and females can't evolve to play a whole spectrum of roles, either within or between species. And yet we still assume that other animals are fairly similar to us when it comes to sex and gender, and we represent those who we know do things differently, like the pregnant male seahorse or the male Australian brush-turkey who guards his eggs, as something exceptional and peculiar.

It's so easy to see the natural world through this lens. Perhaps, instead of talking about males and females, we should talk about producers of large and small sex cells. Sure, it's a bit of a mouthful, but these terms carry fewer assumptions about what the different animals actually do with their lives, and how they organise their reproductive labour. Then again, words are always weighed down by cultural associations – they become loaded. Maybe it's not enough to find new words to describe the similarities between me and a female wolffish – there are few apart from the fact that we both are vertebrates and have teeth, eyes, and produce large sex cells. It's hard to understand that an idea of sex that may feel true to us might not apply to other human beings or other species.

Week 28

When the female Hanuman langur (*Semnopithecus* spp.) gives birth, there are many males who believe the infant is theirs. After all, she has taken the opportunity to mate as widely as possible. It makes the little baby, clinging to its mother's coat as it seeks her teat, safer – she has done what she can to prevent the males around her from killing it. Hanuman langur mothers gladly entrust their young to other females in the group, but if a male comes close and tries to take the baby, it will almost always be to kill it.[196] What's he playing at?

For a long time, no-one could explain the behaviour of this little South Asian monkey, with its long tail, black face and grey-brown fur, renowned for the males' habit of killing newborn young. It was thought that infanticide only happened when there were too many monkeys in the group and not enough food. To scientists, this behaviour was disturbing and unpleasant – surely there can't be any advantage to killing little infant monkeys, the troop's future? But it turned out that for individual male monkeys, there can.

*

The anthropologist and primatologist Sarah Blaffer Hrdy showed in the seventies that the males killed the young in order to mate with the mother. A female will generally not come into oestrus again while nursing, but when she stops – either because the baby dies or because it's old enough and no longer needs milk – she soon releases an egg. A male who kills young he knows are not his own will have an opportunity to mate, thereby passing on his genes at the expense of a rival. New males who come into or take over a group pose a particular threat to newborns. They have nothing to lose and everything to gain from the females going into oestrus and getting pregnant by them rather than expending energy on other males' young. Since Hrdy made her discovery, the same phenomenon has been observed in a number of other species – for example, lions (*Panthera leo*).

But Hrdy didn't stop there. She had found out why male Hanuman langurs were killing their rivals' young, and now she wanted to know how the females responded. It turned out they were far from being mere passive observers of these brutal infanticides: they had strategies for ensuring their young did not get killed. One of them involved teamwork, with the females in a group banding together to chase away strange males, while another was the aforementioned trick of females in oestrus mating with as many males as possible. What's more, Hrdy found that this didn't apply just to males in their own group: females would sneak off and mate with other males nearby, even if this carried a risk in the event that the males from the original group found out. Why did the females do this? A male who has mated with a female cannot be certain the baby isn't his, so there is less chance of him killing it. Female lions do the same thing. They will mate up to a hundred times with males both inside and outside their pride, all in the space of one oestrus period.[197]

When Hrdy published her findings in the seventies, it unleashed a wave of research by female scientists. These women, who themselves encountered challenges and expectations around how they behaved, brought a fresh outlook to the way males and females in various species acted. The white men from privileged backgrounds who dominated science and natural history had discovered many things of value, but there was a dearth of perspectives that could challenge established ways of thinking about sex and gender roles in the animal kingdom. Hrdy's findings paved the way for much of the knowledge we now possess about sex, care, monogamy and evolution, and continue to underpin research into things we didn't think possible about nature. The greater the diversity in scientists' sex, sexuality, ethnicity and class, the more we can question the assumptions we've unwittingly based science on.

The female Hanuman langur has finally come to the end of her long pregnancy. She's mated and carried the foetus inside her, while continuing to mate with as many males as possible in order to guard against infanticide. But her job isn't over yet. Now she can look forward to thirteen months of nursing, even though the baby will start eating solid food at just six weeks old. Luckily, she doesn't do this work alone. She has a support network that consists of her mother and her sisters, aunts, cousins and grandmothers. Everyone helps look after the infant. Only male strangers are a threat. Sisters, aunts and grandmothers take turns to look after the baby, and even today, the first day of its life, the baby might spend half the time in the care of other females.[198]

Week 29

We humans help each other in similar ways. It's the weekend, and there's a schedule clash. My partner has to work, nursery is shut, and the idea of looking after my own child for eight hours straight when I'm nauseous, liable to throw up, regularly have to lie down, can't move my pelvis too much, need to nourish the foetus, to be a vessel for the foetus – it all seems insurmountable. Luckily, my mum comes to the rescue. She sails in full of energy, plans and promises, and trundles off with a child whose eyes are sparkling with anticipation.

I lie on the sofa. I should get up and throw out the withered house plant on the windowsill behind me, but I can't be bothered. There's a limit to how many organisms I can take care of, and I have to prioritise the child I have and the child I will have; the plants are much further down the list. With assistance I can care for both my foetus and the child outside my uterus: I don't need to choose which one to plough my resources into, letting the other die, like the blue-footed booby (*Sula nebouxii*) of the Galapagos islands does when she feeds only the larger chick in her brood and lets the smaller one starve.[199]

*

She's a mystery of evolution, my mother – a female who is no longer of reproductive age. She has gone through the menopause, but she isn't old in human terms. Hopefully she'll live for many years to come. Living beyond reproductive age is a trait she shares with a very small number of other species – just the odd whale and a type of aphid. The fact that it doesn't appear in our closest relatives, the great apes, makes it even more special. Why just us?

The world's oldest identified bird, a Laysan albatross (*Phoebastria immutabilis*) who was ringed in 1956 and later named Wisdom, was still laying eggs in late 2020, and was observed at her regular nesting site in 2022.[200] She was at least seventy-one years old. She's older than my mother and has raised many more young than her, and yet she continues to lay eggs, to pass on her genes via new individuals. Why doesn't my mother do this? Why doesn't she put her resources into producing her own children, rather than playing with her grandchildren and helping her children out? Having grandmothers nearby is not unique to us humans. The special thing is that human grandmothers stop having children themselves. In African bush elephant (*Loxodonta africana*) herds, grandmothers have the highest status – it's the eldest female who leads the way, having gained knowledge over the course of a long time. She knows where to find food, watering holes and safe places. She helps first-time mothers understand how to look after their offspring. She's the knowledge bank in a community where females live in herds with their young, while males go off to live in their own herds as soon as they can take care of themselves. Having an elephant grandmother at the helm leads to more calves surviving and ensures that her daughters give birth to more young.[201] You would think that being herd leader was enough of a job in itself, but she doesn't stop reproducing: an elephant grandmother can keep having young until she dies.[202]

The Japanese aphid *Quadrartus yoshinomiyai* is one of the few creatures who do as we do in this regard: the female stops reproducing long before she dies. Aphids have a complicated life, with generations switching between reproducing sexually and through parthenogenesis – unfertilised reproduction, as seen in the water flea and the Komodo dragon. The whole thing starts with a female aphid with wings overwintering on her chosen host plant, a bush belonging to the witch hazel family. When spring comes, she chooses a spot in which to start her colony. Perched on her favoured branch, she will manipulate the host plant's tissue to form a gall – a swollen part of the stem or leaf. She gets comfortable inside the gall, and then begins to birth live young. She delivers between fifty and two hundred daughters, produced through parthenogenesis. These daughters will live their whole lives inside the gall, which is why they have no wings.

In time, they'll produce their own daughters, also through parthenogenesis. But *these* young, the grandchildren of the first female aphid, will have wings. This new generation lives for a long time inside the sealed gall, which has no opening and is therefore well protected against predators. But when they're ready to come out into the world, the gall needs to be opened. At this point, the wingless mothers stop reproducing, and, for the next few weeks, until all their winged daughters have left, they will sit by the gall's opening, producing a wax-like substance from glands on their abdomen. If a predator – a ladybird larva, say – tries to enter the gall, the wingless aphids shoot this wax at the intruder. The wax solidifies, and the ladybird larva is scared off – with a kamikaze wingless aphid still attached to it. By living long enough to defend the colony and give more of the winged aphids the chance to fly out and create new colonies, the post-reproductive, wingless aphid ensures that more of her own genes will be spread.[203] The difference between these aphids and us is that the aphid's post-reproductive contribution is directed at the generation directly after her, her own offspring. The human grandmother's contribution is directed at her grandchildren.

The fact that my mum comes to pick up the three-year-old, giving me a break for a few hours, has little bearing on whether or not my child will survive. I probably would have managed being home alone with both the foetus and the three-year-old, even if that meant there was a little more kid's TV during the course of the day than I would have liked. Norway has one of the lowest levels of infant mortality in the world,[204] and studies seeking to measure whether having a grandmother present improves a grandchild's life chances in wealthy, industrialised nations show varying results.[205] But the knowledge that we had grandparents close by who wanted to help us was critical in my decision to get pregnant again, helping me come to terms with the prospect of spending so many months feeling nauseous and vomiting as I had last time. My mother, walking out the door with her delighted granddaughter, is the modern version of generations of grandmothers who've ensured that their children have more children, and that the grandchildren have a greater chance of surviving to become parents themselves.

Anthropological studies of modern hunter-gatherer societies, who live as close as possible to the way we think humans lived before the advent of farming ten thousand years ago, have given rise to the "grandmother hypothesis", which might explain why people live so long after their reproductive age. It posits that women who stop reproducing themselves, but instead focus on helping their children and grandchildren, increase the chance of their own genetic material being passed on. The anthropologist Kristen Hawkes, now a professor at the University of Utah, has studied the role of older women in groups such as the Hadza people in Tanzania. Together with mathematical models, observations from contemporary hunter-gatherer groups give strong indications that the menopause evolved because helping one's own children to raise their infants, both by gathering food for them and looking after them so that the mother can use her time to either gather food or produce more children, results in healthier grandchildren who are quicker to eat food other than their mother's milk, which in turn leads to their mother becoming fertile more quickly and having children at shorter intervals.[206]

*

But there is still debate as to why the females of a few species live so long after they've stopped reproducing.[207] The grandmother hypothesis could, as stated, show that there's an advantage to having a living grandmother, and we can imagine that those whose grandmothers were still alive got a tiny evolutionary boost that has spread over time. But it's difficult to pinpoint exactly which mechanisms and functions combined to ensure that grandmothers, or the phenomenon of living far beyond the age at which you can reproduce, came about in humans and those few other species.[208] Perhaps it's not so difficult to see that it might be nice to have a grandmother at hand. But how is it advantageous for a grandmother not to have more of her own children in addition to looking after those she already has and their offspring? From a purely genetic perspective, she is closer to her children than her grandchildren, and for the biological calculation to add up, there would have to be greater advantages in ceasing to reproduce and assisting with her children's reproductive efforts instead.

Perhaps we can find the answer by turning to a species in which grandmothers take a leading role. Orcas (*Orcinus orca*) are found in a variety of marine environments, but we know most about those which inhabit the waters off the west coast of the USA and Canada.[209] Orcas live in large groups – or pods – led by a matriarch, an older female with life experience and knowledge. Like us, they can live for many decades after they have stopped reproducing. The pods that live near the Pacific coast of North America like to eat Chinook salmon, a species that can vary in number significantly from one year to the next. The quantity of salmon available in any given year has a major impact on how many orcas survive, and how many reproduce that season. With their long experience of the salmon's movements and potential whereabouts, grandmother orcas have immense value to the pod, especially in lean years – and the consequences of losing a grandmother are much more severe at times when there are fewer salmon available.[210]

These orca pods have an unusual family structure, consisting of the senior female and all her offspring, both sons and daughters. No-one gets cast out when they reach reproductive age, and no new members are admitted. To prevent inbreeding, members of the pod take the opportunity to mate with their peers in other pods when they meet.

This means that the older a female is, the more closely related she is to those in her pod. A newborn female has her mother and siblings, but since her siblings probably have different fathers, they are less closely related in genetic terms than a mother is to her own offspring. The older a female gets, and the more offspring she has in her pod, the greater her in-pod kinship. At the same time, research has shown that when mothers and daughters produce young at the same time in these groups – that is, when grandmothers have not yet stopped reproducing – the mortality rate of the grandmother's calves is much higher than that of the daughter's.[211] Evolutionary theory states that a female with low levels of kinship to others in the group will compete harder for food and any other resources she needs to reproduce, while a female with greater kinship in the group will be less incentivised to do so, because her children are competing with her grandchildren, who also carry her genes. She is in so-called reproductive

conflict with younger females – a conflict over whose offspring will get to grow up – but she is at a disadvantage because she cannot just seek out food and resources for herself and a single calf: in other words, she faces higher reproductive costs. At a certain point, a female orca will have such close kinship to her whole pod that it makes sense, from an evolutionary perspective, for her to focus on helping her sons and daughters, rather than continuing to reproduce.

Orca grandmothers have their sons in their pod all their lives. This gives them an additional reproductive advantage than would be the case if they only had daughters nearby. Because males reproduce too, they also spread their genes when pods meet to mate. However, the cost of this gene-spreading, in the form of the additional food needed by pregnant and nursing mothers, will fall to the other pod. Every time a grandmother ensures that her son gets enough food to grow to maturity, and then leads her pod to other pods so he can mate, she is helping to ensure her genes are passed on, at the lowest possible cost. This might explain why orca grandmothers, but not elephants (where males leave the group), stop reproducing at a certain point in their lives – the elephants can continue to reproduce because they are no longer responsible for looking after their male offspring.

So, what about humans, who don't live in pods or herds led by a grandmother? There is much to indicate that in the earliest human societies, young females who reached reproductive age were the ones who left a group to join a new one. And, in contrast to orcas, the earliest humans probably mated within groups.[212] If that is the case, it brings about a situation where a young female who has just switched groups has low kinship with those she lives with, giving her an advantage over older females living with their children and grandchildren, who will have a high level of kinship with the group as a whole. This means we have the same reproductive conflict as the orcas, which might explain why we're two of the few species that have developed both a key role for grandmothers and the menopause.

My mother, the orca grandmother, the male titi monkey, the Hanuman langur aunties, the Damaraland mole-rats, the male emperor penguin and seahorse, the un-paired African social spider, and the older beaver siblings are all helpers who do their bit to ensure that the young in their group grow up. They are just a few of the many examples of species in which it's not only the ones with the eggs who are responsible for making sure that their group's demanding offspring have a chance in life.

Week 30

From now on, my baby can be born without too many major complications. It would be much too early, but we're long past the weeks when I kept expecting to see a patch of blood in my knickers, past the weeks when all hope would have been lost if something had gone wrong, past the weeks when the baby might perhaps have been saved had I gone into labour, but most likely with serious consequences. We've now entered relatively safe terrain. Hopefully, the foetus will remain inside me for another ten weeks, growing longer, putting on the fat that will protect him from the cold of the outside world, reaching a size that will make him more robust in his encounters with his surroundings.

As I start to feel more and more confident that my pregnancy will go well, that I'll come out on the other side with a lovely, live baby, the need grows to make everything ready for his arrival. I haul out the old pram from the basement, wash baby clothes that are still much too large for the little foetus and make lists of what needs to be done before he arrives.

Because suddenly it feels absolutely necessary that the

kitchen units are scrubbed clean, that those old papers in the recesses of the box room get sorted, and that the cot is carried down from the shared attic and into the bedroom, even if I'm planning to have him in bed with me to start with.

Is it the animal in me that feels an urge to act like the beaver, gathering leaves for its little kits to sleep on; like the rabbit, who pulls out its fur to make a soft nest for its young; or like the Nile crocodile, who spends a long time finding the very best nesting place? Is it a biological instinct kicking in to make me get the house ready so that the new baby will have the best start in life? Or is it society's expectations of me as a mother that makes me think it's important?

Nesting, whether in birds who build real nests, or rabbits who dig burrows and pull out fur, is essential for the survival of the offspring of many, many species. Not even animals who reproduce by releasing eggs and sperm into the water do it at any old time – they coordinate spawning to guarantee the highest possibility of the eggs and sperm meeting, while the danger of sex cells being eaten by predators is reduced when many individuals release them simultaneously.[213] Building a nest or making the surroundings as safe as possible to ensure the survival of one's offspring is an adaptive behaviour: those who have, throughout evolutionary history, done a tiny bit more than others to prepare, have had a slight advantage, leading to more young surviving, and so the behaviour has been passed on.

Not much research has been done into nesting in humans, even though our tendencies are often described as a natural instinct on websites aimed at pregnant people. But one of the few studies that have been conducted shows that pregnant women do what other animals do: we prepare our home for the incoming family member. This can mean anything from buying a pram and painting the nursery, to wiping down cabinets and checking all the baby clothes are clean. Studies have also shown that pregnant women avoid unfamiliar places in the period leading up to the birth, and prefer to be with people they're close to, be that friends or family.[214] That makes sense – keeping clear of strange places means avoiding potentially dangerous situations and unfamiliar bacteria, and reinforcing social bonds with family and friends can give the baby social advantages in the form of more people who want to look after them.

But this research met with resistance. Can we really differentiate between the influence of culture and biology in these matters? It's still the case that around the world, women do more housework than men, even in apparently more equal countries such as Norway.[215] And the tasks often held to be evidence of nesting in humans, such as tidying and cleaning, are among the most gendered aspects of housework: in heterosexual households they are mostly undertaken by women. At the same time, women are judged more harshly than the men they live with if a house seems untidy or dirty.[216] It's hard to distinguish between biological instincts and cultural practices so deeply ingrained that they feel instinctive. Neither is there any certainty of being able to tell biology from culture when we look at isolated phenomena, because biological and cultural influences are closely intertwined. Something as simple as your height is biologically determined by the genes you inherit. But height is also affected by which foods and how much of them your mother got when she was pregnant with you, which foods you ate as a child, what climate you grew up in, and what illnesses you were exposed to. Not everything is biology, and with something as complex as human behaviour in contemporary society, it's extremely difficult to determine which factors are interacting to ensure a particular outcome.[217]

Regardless of whether the reasons for me preparing my home for the new family member are biological or cultural, it's practical to have some clothes, a pram and a bed ready for the baby's arrival. Perhaps the reason why I'm doing what I'm doing now is as simple as that – it's practical – even if it feels like the need comes from somewhere deeper within, an urge to get it done so I can go on brooding in peace.

We carry down the cot. The three-year-old fills it with cuddly toys the baby can inherit from her. I fill the chest of drawers with freshly washed baby clothes, neatly folded into little piles. We're ready.

Week 31

By this point in the year, the only reindeer who has large antlers is the female carrying a calf. Throughout the winter she's been using them to show her dominance, to mark her place in the herd's hierarchy and to secure food for herself and her yearling. The larger her antlers, the higher her rank, and the greater access she has to the areas with the best food. But now she's chased away last year's calf, and it has to fend for itself, because she's about to give birth to the one she's been carrying throughout this long, cold winter.

Reindeer (*Rangifer tarandus*) live a nomadic life. They're constantly on the move, from their lichen-draped winter feeding grounds to the areas where the grass first germinates in the spring. In the summer, they roam plant-rich grazing grounds, and there the female will get the nutrients her body needs to recover from her previous pregnancy and produce milk for the calf, before the herd moves on to its autumn haunts, where there are mineral- and nutrient-rich fungi. This species of deer is specially adapted to life in the open expanses of the frozen mountains. Their hair fibres are hollow and full of air, insulating their bodies like a puffer jacket against the harsh winter winds, and helping them to float as they swim across fast-flowing rivers. As the herd moves about, knowledge is passed from mother to calf – about where to find food in each season, where it's safe to give birth, where mating takes place.

The reindeer – the iconic species the first inhabitants of the Nordic region depended on for hunting, the one they followed northwards after the last ice age, the one we have domesticated and still follow nomadically in Northern Scandinavia – ranges freely across mountainous regions of Norway to this day. Within the soft, furry muzzle, air enters a special conch-shaped organ which warms it before it's drawn down into the lungs and then retains this warmth when it's exhaled.[218] This makes for excellent energy efficiency – which is helpful given that winter temperatures can reach forty degrees below zero. The reindeer is the only deer species in which the female (the cow) has antlers; it's her means of protecting her calf, the way she ensures she herself gets enough food to build her foetus, muscle by muscle and bone by bone. The male loses his antlers in the autumn, after the rutting period, when he's done fighting other males to win the females' favour. The bigger and more elaborate his antlers, the more attractive he is to her.

If a cow doesn't get pregnant in the autumn, or the foetus miscarries, she loses her antlers some time before calving season. As winter relaxes its grip and fresh green shoots start to burst up out of the ground, competition for food is no longer so tough, and losing her antlers early can mean her new pair come through quicker – allowing them to grow larger next winter and giving her better grazing chances next year. Perhaps then she'll manage to hold on to the foetus, so it can become a calf.

The cow, antlers still in place, is about to give birth, two hundred and twenty-five days after her short oestrus in the autumn. Throughout the long winter, she has fought to secure the best grazing and find food herself and her family, tilting her head slightly to announce her dominance and chasing away cows with smaller antlers. And she has survived, with one calf at her side and one in her uterus, but now she's focusing only on the one inside her – the other must look after itself. She moves away from the herd to give birth in peace, and to be able to fully pick up the calf's scent, as it stands on unsteady legs. After an hour at most, it suckles for the first time. In just a few days it will need to be ready to follow the herd onwards, ever onwards, in the eternal, cyclical quest for food.[219] [220]

Throughout the cow's reproductive life, as she passes on her genes through calf after calf, there is another life cycle taking place in her skin. The reindeer botfly (*Hypoderma tarandi*) is a large fly with yellow and orange fur that looks a little like a bumblebee. But it doesn't lay its eggs in a nest, like bumblebees do. The larvae of the reindeer botfly need reindeer flesh if they are to survive. In July and August the female botfly lays her eggs in the reindeer's fur, right up close to the skin. The eggs hatch, and the little larvae burrow down and eat their way beneath her skin until they reach a certain point on her lower back. At the very end of autumn, the larvae encapsulate themselves in connective tissue with just a tiny air hole out through the reindeer's pelt. The reindeer carries her foetus in her uterus and the botfly larvae in small cavities in her back, like the Suriname toad, only, in the case of the botfly at least, she's an involuntary host. After she has delivered her calf, the reindeer botfly larvae crawl out through their air holes and drop to the ground, where they pupate and wait for June to come so they can emerge, fully grown, to mate, lay eggs on new reindeer, and die.[221]

I'm standing up on the bus, poking my belly out of my coat in the hope that someone will get up and offer me their seat. I'm beginning to develop pelvic pain, and I can't stand all the way home – if I do, the discomfort will get much worse. But it's tiring to have to keep asking to sit down; the middle-aged men with headphones clamped to their ears don't want to look up and catch my eye. They can't be bothered to stand up themselves, they're hoping someone else will offer me a seat. I wish I had antlers, so I could dominate these antler-less, pregnancy-less males who've done their day's work and don't need the extra resources at this time of year.

Week 32

My pelvis is aching more and more. At the front, under my growing belly, the pain beams down towards my groin. If I'm able to stay still it improves, so I try to avoid twisting or carrying anything heavy, but I live on the third floor, with a child who's almost four, who regularly has temper tantrums on the way up. She's sooooo tired, and cries ring out in the stairwell. I know my neighbours can hear every sound, they can hear that she's refusing to move, hear me start by chiding her quietly, before offering treats in return for her walking up by herself. Every day I have to offer more sweets to get the same effect. It's a vicious cycle, but we get up the stairs and in through the door, and the moment it closes behinds us, the mood lifts, because now it's time for the reward I've promised. I eat a few jelly babies myself – I could do with the sugar.

I know that every step on those stairs will lead to greater pelvic pain later, when I'm on the sofa with a cushion between my legs. Ideally, I would prefer to carry just myself and the foetus up to our apartment, and yet I find myself having to carry the feelings of a soon-to-be-four-year-old too.

The bones that form my pelvis have to move a few millimetres apart in order to accommodate the baby's head as it emerges, rotating its way along my narrow birth canal. For a long time we believed that the human birth canal was narrow as a compromise between walking efficiently on two legs and delivering our babies, and that this was why men have smaller pelvises than women. But we humans are not the only species in which the size of the pelvis varies between females and males, with the male's being smaller even if he does not walk on two legs.

A female chimpanzee (*Pan troglodytes*) is about to give birth. The baby's head only fills about 70 per cent of the narrowest part of her birth canal, and yet the foetus rotates through the channel from the uterus and out into the world in a similar, if not identical, manner to how my baby will in two months' time. A chimpanzee birth is easier than a human birth, but even though the baby's head doesn't completely fill the birth canal, male and female chimpanzees still have different pelvises, just like us. The same is true of a number of other species. Even in the case of the Virginia opossum, whose newborn young are only 0.01 per cent of the mother's body weight, the male and the female have slightly differently shaped pelvises, regardless of the size of the adult animals.[222]

Perhaps the pelvis did not become narrow to make it easier for us to walk upright, despite our long-held assumption. After all, female humans walk and run just as efficiently as males, even though they have a wider pelvis. A new hypothesis, which has held up when tested in experiments, is that the pelvis narrowed so that our pelvic floor muscles could hold in our intestines. The muscles of the pelvic floor are unable to stretch far – they are attached to the various bones that comprise the pelvis. Stretching the muscles across a wider pelvis can cause problems: for instance, female humans with wider pelvises are more likely to experience prolapse and incontinence.[223][224] Meanwhile, Geoffroy's tailless bat, which never walks upright on two legs, and spends most of its time hanging upside down, doesn't need pelvic floor muscles to hold in its intestines in the same way as I do. The females can afford to have a pelvis that allows her to deliver her offspring with comparative ease, even though it can reach up to 45 per cent of her own body weight.

My pelvis is robust, but my hormones are causing the ligaments holding it together to expand: my body is preparing to give birth to a baby with a big, big head. The symphysis, where my pelvic bones meet at the front under my large belly, is flexible – an evolutionary adaptation that makes it easier to give birth – but not as flexible as the guinea pig's. On average, the human symphysis expands by three millimetres, in addition to the pelvis expanding a little at the two joints by the spine. Why is my pelvis so inflexible, when the guinea pig is able to make birth so easy for itself? It all comes down to the fact that too much flexibility can make the pelvic floor weaker, and intestines that fall out hardly offer an evolutionary advantage.[225]

*

This sex-determined difference in pelvis size exists in all large placental and marsupial mammal groups, and in certain reptile species too, leading some scientists to reason that it was present in early mammals, or perhaps earlier still, in the common ancestors of reptiles, birds and mammals.[226] It may have arisen to make it easier to deliver offspring or eggs that were large in comparison with a female's own body size, and the fact that we see it in animals with tiny newborns could mean it's a left-over historical trait for them, one evolution hasn't found it necessary to do away with.[227]

Week 33

When the common ancestor of myself and the hippopotamus (*Hippopotamus amphibius*) began to come up onto land around 375 million years ago,[228] it needed a way to keep its foetuses moist without having to return to the water. It already had some form of lungs to breathe with, and limbs for getting around, but to avoid being dependent on water when it came to laying eggs, like frogs and salamanders and other amphibians are today, it had to innovate. At least 318 million years ago, our eggs developed membranes that stopped them drying out, protecting the little water-borne foetus from the dry air on land.[229] But that wasn't enough for evolution; we went from laying eggs to keeping them inside us, looking after them, protecting them. We carried the primordial sea within us as we came up from the shore onto the beach, into the forest, into the trees. Some creatures thought land was overrated, they wanted to get back to the water, and these whale and hippo ancestors did just that, become water-loving animals who keep their foetuses inside them. While the whale went all in and committed to the oceans, the hippo stayed at the

water's edge. It's semi-aquatic, needing water to live, both as a foetus and an adult, but it only eats plants that grow on land, so it has to leave the water to find food. When the hippo gives birth, it does so in the same way as the whale, with the young coming out hind legs and tail first, because it needs to swim to the surface to breathe. If it were to get stuck in the birth canal with its head out and its lungs under water, it would die. It needs air as it transitions from the amniotic fluid in the uterus to the water outside.[230]

Over the thirty-three weeks of her pregnancy[231], a female hippopotamus gets heavier and heavier, but the water supports her, making her body weightless. I think of her the one time I take the three-year-old to the swimming pool. I'm nauseous from the sudden sensation: it's as if my body had at last found an acceptable way to be pregnant, and the weight is an important part of that. Once I'm in the water, my intestines start to reorganise themselves again. It's like being in a tiny plane that's sent into free-fall for a moment by turbulence, or the on downhill parts of a roller coaster, except that instead of lasting thirty seconds, it persists for the whole hour my three-year-old insists on being in the splash pool. I sit on the edge, but keep getting cold and having to dip my body into the warm weightlessness once more, slipping in and out of the water again and again.

The hippopotamus is enormous – the third biggest land mammal after the elephant and the rhino. Its body is round, its legs short, its head broad, with bushy ears on an otherwise rather hairless body. Its wide snout has bristles and enormous canines. During the day, the hippo lives in fresh water, cooling itself in the lakes and rivers of southern Africa. At night they can walk several kilometres in search of food – they exist on the boundary of land and water.[232]

When the time comes for her to give birth, the female leaves her herd to deliver her calf in peace, giving it a chance to get to know her before they return to the others. She can give birth both on land and in the water – she's an in-between creature in so many ways – but wherever she decides to do it, the calf comes out hind legs and tail first. It then fills its lungs with air for the first time, before beginning to suckle. She nurses it in the water, because that enables her to lie on her side, and the calf closes its nostrils in order to feed without getting water up its nose.[233]

Week 34

It's getting milder outside, spring is on the way. In less than two months I'll be pushing a pram in the sun, looking at the flowers gradually coming to life, the trees sprouting tiny, pale green leaves, the first bumblebees searching for a safe place to build a nest. I'll move my heavy load from my belly to the pram, my body will be lighter and yet simultaneously weighed down by the lack of sleep and the hunger that follows when my nipples are the only source of food for my growing baby. It's still cold out, but the sun is ever higher in the sky, the evenings are lighter, and the rays from our great star bring with them the promise of better times.

Inside a deep burrow on an Indonesian island, the abandoned eggs of our Komodo dragon are hatching.[234] The mother, an enormous lizard that can choose whether or not she needs a male to reproduce, has long since moved on. When the rainy season began, about sixteen or seventeen weeks after she laid the eggs, she stopped guarding the nest. Over those long weeks, she got thinner and thinner. Eventually she had

to make a choice: leave the nest before the eggs hatch – or die. One theory is that the eggs have out-evolved her: they have developed a longer incubation period, but the female hasn't evolved to stick around for the duration. It's possible the nest doesn't need guarding after the rainy season has set in: perhaps there are fewer predators then. Or perhaps she makes herself scarce because Komodo dragons are cannibals, and her hunger would have taken over once her young started crawling out of their eggs. The strong instinct that led her to watch over the nest for so long would have been overcome by her need to survive, and the first meeting of a mother and her young would have ended in a bloodbath. Regardless of the reason, she's long gone, and the baby dragons crawl quickly out of the hole and up into the trees. They hatch towards the end of the rainy period, when the food they live on in those first weeks – insects – is abundant. They're on their own now. If they're not careful, they might get eaten by adult Komodo dragons. Luckily, the adults are too big and heavy to climb trees.[235]

Week 35

I catch myself talking to the foetus inside me more and more. My doctor has written me another sick note, I'm home alone all day, catching up on the sleep I lost getting up to pee at all hours of the night. I'm resting ahead of my shift with the three-year-old once she's back from nursery. I mostly talk to the dog, who sits loyally beside me on the sofa, but perhaps the foetus can follow what I'm saying, though if he doesn't understand the words, maybe it makes no difference whether I'm speaking to him or to the dog. I think my voice is probably reassuring to him, I think he can recognise it as it flows through my body, through the amniotic fluid that surrounds him, and into his little ears.

A friend comes round with her three-week-old baby, and when she's putting on her sling, I hold the little one. The baby starts getting upset. I talk in my baby language, chatting away, but it doesn't help. But then her mother starts talking, and it's obvious that the baby recognises her voice – she stops flinging her head about and listens, drawing in her outstretched arms and growing calm. She hears that her mother is nearby, and that comforts her; now she can stay with me while her mother is getting ready to go. Then she's placed into the carrier, right at her mother's breast – perhaps it makes her feel as though she's back in the womb, pleasantly and cosily constricted, with her mother's voice vibrating through her body. This baby is content.

I'm not the only one who talks to their foetus. The common bottlenose dolphin (*Tursiops truncatus*) is pregnant longer than me – almost a whole year.[236] Each bottlenose dolphin has its own signature sound, which they use to talk to others in their pod: they call "Here I am", and the others respond.[237] A young dolphin's own signature sound, which it chooses for itself once it's old enough to make adult sounds, is different from its mother's. It imitates those of distant acquaintances, to create a "name" that's easy to distinguish from those of the others in its pod.[238] When the dolphin is still a foetus, the mother dolphin calls her own name more often,[239] presumably so that the calf will learn it and come to her when she calls it after it's born. Her young quickly learns to swim well, and it's impossible for her to put reins on it, but it must learn to come when she calls. If it confuses her with another dolphin, it might become separated from her and die.

Newborn humans prefer the voice of the person who gave birth to them. They turn towards the sound and are calmed by it. Human foetuses in weeks thirty-six to forty-one can distinguish between different sounds and different languages, and foetuses that are almost at term can distinguish between male and female voices.²⁴⁰ Both my baby and the dolphin's calf will depend on their mothers for their survival. Recognising our voices among all other sounds is the first step in finding us, coming to us – a lifeline after the umbilical cord has been cut.

Week 36

I'll be on display when I give birth. In a strange room I may or may not have been in before, where unfamiliar people could enter at any moment, in a large building in a city a long way from where I grew up, probably with a midwife I've never met before. In this city there are trees, trams, parks and tall buildings, rather than brushwood and tufted hairgrass, reindeer on the other side of the garden fence and rows of little wooden houses, like where I was raised. But I know this place, I've habituated myself, become accustomed to my surroundings; I navigate like a native. I've chosen this city and this hospital. I've asked for the opportunity to give birth on this ward, where I might have met some of the midwives before. I could have chosen to give birth at home if I'd wanted. But the grunts and yells of my labour feel like too much to subject my neighbours to, and my bathroom's a little too small for a bath. I want the public hospital's enormous birthing pool, their special issue red berry squash, the security of knowing that there are multiple midwives on hand.

*

In a zoo in another country, a gorilla (*Gorilla gorilla*) is flat on her back. She's bigger than me, but will give birth to a smaller baby after having been pregnant almost the same length of time.[241] She's not in the jungle, so in theory she's out of her element – but maybe she was born here, maybe the zoo is all she's ever known.

Her belly was big to begin with – she's a herbivore, and the large stomach provides space for the digestion of fibrous leaves – but now it's larger still. In the wild, when her time comes, the female will distance herself a little from the rest of the troop during one of their daily pauses while gathering food. Gorillas stay close to the protection offered by the troop, but they manage their labour on their own, without help; it's over in thirty minutes.[242]

The gorilla in the zoo is holding on to her feet, the thumbs on her hands against the thumbs on her feet, as she pushes out the baby. When the head is just out, she feels it with her hand. The face is turned up towards her, and she puts her hand behind the baby's neck, supporting the head through the next contraction, as the whole body comes out. She's in her home, in the zoo, giving birth on a bed of hay behind a trellis, rather than the vegetation of the wild. The staff on the other side of the fence sigh with relief when the baby is delivered. The other gorillas are sitting on the far side of the enclosure – this is her job to do alone, but they are close at hand. She sits up, holds the baby's head with both hands and supports its body with the long toes and thumb of one of her feet, gazing into its eyes and licking off the mucus covering its nose. Now there are two of them and they'll stay close together. She nestles her baby close to her body, where it will soon start holding on to her fur as it searches for her nipple, and with it, food.[243]

We humans are the only creatures who do not give birth without help. Our pelvises may expand a few millimetres during pregnancy, but it's still an extremely tight fit, and the foetus must rotate along the birth canal and lead with the back of its head if it's to make it safely out. And that's not all – the channel is not the same shape all the way along. At the top, where the baby's head will start its journey, the passage is widest sideways, if you're looking at me head-on. Halfway down, it's widest in the other direction, between my belly and my back. The foetus has to turn in order for its big head and broad shoulders to get through, and he comes out with his head in the opposite direction to the gorilla baby's, with his face turned away from mine. If I were to try to hold his head myself, without the help of a midwife, I would risk damaging his delicate neck by bending it backwards.[244] The average diameter of the human birth canal is thirteen centimetres at the widest point, and ten centimetres at the narrowest. The average diameter of a newborn baby's head is ten centimetres at the widest point, and the distance from shoulder to shoulder is twelve centimetres.[245] In other words, it's a tight fit.

In theory, the human female can give birth alone. But we do so extremely rarely. Anthropological studies have shown that in a very few groups there are some births that take place without help from others – for instance, a mother's first will take place with assistance, but the others after that will be solitary. Still, the vast majority of births in the vast majority of human cultures take place with the help of a partner, an experienced birthing companion in the tribe, or others with medical competence.[246]

It's probably nothing new. Our large head and narrow

birth canal have evolved simultaneously with greater cooperation and assisted births. Humans are not the only ones with a narrow birth canal and a large head. For example, still births have been observed in some great apes as a result of the baby's head being too large for the birth canal. But apes do still have the advantage of the baby being born with its head turned towards the mother, even if the baby has to twist its way through the birth canal a bit on its way out.[247]

We humans actively seek help when we are in labour, and assisted births have been documented since the early stone age,[248] though they've probably been taking place for much longer than that. When your child is born facing away from you, it's not only hard to support the head on your own, it's difficult to clear the mucus from its mouth or remove the umbilical cord that might have got wrapped around its neck, and you're not capable of getting the baby's shoulders into the right position if it gets stuck. You need help.

Our difficult births have evolved as a result of various human traits. We have a narrow pelvis to keep our intestines in place. We have gradually developed bigger and bigger brains, and therefore a larger head at birth. We collaborate. Some scientists think that getting help with births was essential for us even when we were first beginning to walk upright, between 4 and 5 million years ago.[249] [250]

Gradually, our bodies and our behaviour have evolved together. Getting help with labour meant that fewer labouring mothers and fewer newborns died. And this has probably impacted the evolutionary pressure on the rigid pelvis and the narrow birth canal.[251] It may be precisely because we get help that it is so narrow – those who have asked for help during labour have fared a little better, and so this behaviour spread as the human species evolved.[252]

But perhaps it's not just physical assistance we've evolved a need for during labour. Humans are a social species. We come together when we are happy and when we're sad, we ask for help when we're in pain, we need the comfort of others when we're afraid. Perhaps this tendency to ask for help when we're sick and in pain, to seek solace and warmth from others when we're afraid, is also a need and a behaviour that has evolved as our bodies and our deliveries slowly changed. We need comfort and care. We give birth with helpers around us because getting physical help improves our chances of survival, but at the same time, we don't give birth alone because we've evolved a need for emotional support and security.[253]

I won't give birth alone: I'm in safe hands. But modern hospitals and modern medicine can be a threat to me and my baby. Modern birth support, medical knowledge and medicine have saved the lives of millions of people. Without them, we would have much higher death rates in mothers and infants. But hospitals haven't always been the best places to be for women in late pregnancy. In the mid-nineteenth century the number of women who died of puerperal sepsis in hospitals across Europe was extremely high – up to 25–30 per cent. Those who could afford it gave birth at home, where the risk of this postpartum infection was much lower. It was the very poorest women, prostitutes, and unmarried mothers, who had to give birth in hospitals. The doctor Ignaz Semmelweis discovered that the death rate at a hospital in Vienna was much higher in one of the delivery wards – the one in which the medical students were instructed; the other was reserved for the practical training of student midwives. And the difference between the doctors and the midwives was that the medical students came straight from the dissecting rooms, where they'd been cutting up dead bodies, without washing their hands before sticking them inside the labouring women.

When Semmelweis made sure everyone started washing their hands, the rate of infection dropped dramatically, and yet it still took a long time for the practice to spread across Europe. Many people were sceptical of Semmelweis's discovery, and many women died needlessly before all clinicians started washing their hands properly.[254][255] Now everyone does – the infection rate is low, and few die. Thankfully, modern medicine saves more and more people who would otherwise have died or become seriously ill. For instance, today we can check all newborns for a range of conditions including Phenylketonuria (PKU)[256], a metabolic disorder that can lead to brain damage if left untreated. All that's required is a small blood sample, and children who have this genetic mutation can be treated with a special diet that leads to them developing healthily.

Things are infinitely better than in Semmelweis's day, but even quite recent delivery care has lacked the knowledge needed to give the best start to newborns and their mothers.

For example, in many Western countries during the 1950s and 60s, there was an extended lying-in period. Hospitals had taken over from traditional home births, and in so doing, they also took responsibility for the postnatal period during which the mother traditionally rested. The lying-in time was at least eight days. But as the onus was on the mother's recovery, she was only allowed to see her baby every four hours. That was how often people thought the baby should be fed, and they saw no need for care and closeness in between.[257] Instead of playing with her child, the mother should rest. It was actually believed that newborns had no needs over and above their hunger. It was thought that they experienced neither pain, fear, nor loneliness, and descriptions have been found of heart operations being performed on newborns completely without anaesthesia, right up until the 1980s.[258] Why use anaesthetics on a little body that couldn't feel pain? These wordless little people must have experienced great suffering. Now we know that all humans experience pain, and that babies need to be near a caregiver, that they need contact with other people in order to develop fully.[259] Today, babies in Norwegian hospitals are put to their mother's breast immediately after birth, and the focus is on the mother and baby establishing a strong bond right from the beginning, through skin and eye contact.

However, even as we have begun to understand that our babies are living, feeling creatures who need to be close to us, who need to smell and hear us to feel calm, it seems we've forgotten that the bodies of those of us who give birth are the result of millions of years of evolution. Our deliveries are the product of the evolution of the human pelvis and brain, and of the habit of collaboration and assistance around birth – both physical assistance and the emotional help we gain from having familiar, supportive people close at hand. Fear and stress make contractions worse: the body won't want to give birth if it doesn't feel safe. Some scientists think that it was fear and pain, and not the knowledge that our birth canal was narrow, that led us to seek help with birthing in the early days of the human species. And since then, throughout the history of human birth, this emotional support has been as important as physical assistance.[260]

But is due attention given to that now? The time we spend in hospital has been cut dramatically, and assumptions have changed as to how quickly we'll be able to manage by ourselves after labour. We go into hospital expecting a safe and supportive delivery, but hospitals are no longer geared towards helping us cope with evolutionary fear. Labour often takes longer than a midwife's time on shift, and there's no certainty that a hospital will have enough midwives available for one to be with you at all times.

An article reviewing research in this field showed that having one person present throughout labour – something that is now the exception rather than the rule in hospitals – can make labour better for both the person giving birth and the baby being born. The review demonstrated that when the same person was present throughout, it resulted in a higher chance of birth without medical induction, a quicker delivery, and lower rates of caesarian section and other interventions.[261] Yes, you're probably thinking, but most of those people have their partner with them throughout the birth, or a friend or a parent. And it's true that it's important to have family and people close to you at your side. But the research review showed specifically that it is best to have continuous support throughout labour from an individual outside the birthing person's family with prior experience of assisting births, an individual who is there solely to support the person giving birth and has no other duties in the hospital. An individual who has other duties may become stressed trying to deal with everything, and family, partners or friends generally have little knowledge of the birthing process. They might themselves require support when they see someone they love going through labour, or they might be afraid that something will happen to the baby during the birth.

That individual – someone with experience of labour who can be present without having other responsibilities – is often called a doula. They support the person giving birth without having any medical responsibility or tasks.[262] The benefits of doulas are gaining wider recognition. The Norwegian health service, for instance, now offers multicultural doulas to women who have recently come to Norway and whose living situation is precarious.[263]

But getting the continuous support that a doula or a midwife with ample time on their shift can give, is still the exception rather than the rule in most hospitals. To stand a chance of benefitting, you need to give birth during low season, and in such a short time that the midwife who's there when you arrive doesn't go off duty while you're in the middle of labour. That's pretty unusual.

*

Modern medical practices change when we make new discoveries. But we have to see it to believe it, just as with the absence of sexual monogamy in birds. It took a long time for Semmelweis's realisation about handwashing to catch on, because his critique of existing procedures was considered provocative.[264] So what about the realisation that those giving birth need emotional support, not just a healthy baby and a relatively intact vagina?

When I'm about to give birth, a note is made of whether the baby is to be born by caesarian section or with the help of some other intervention, and whether I'm to have anaesthetics or medication to induce labour. State hospitals are financed by the government on the basis of the treatments they offer; maternity departments are funded on the basis of how many births take place there, and what type of medical treatments are administered during the birth. But this system doesn't take into account whether it was possible for staff to give me continuous care throughout. There is no record of the fact that I was so afraid my contractions stopped, or of the comfort and support given to me by an experienced midwife who was able to get me through it without the need for a doctor to be called. The system doesn't consider breastfeeding guidance as treatment, nor can it highlight the socio-economic benefits of new parents who go home safe in the belief that they are going to be able to give their child love and care, and don't need to contact their GP or other medical services so often.[265]

Maybe the idea that care is important, even if we can't put a price on it, and the understanding that our bodies need not just medical help, but also comfort and security, will be modern medicine's next discovery?

Week 37

My stomach stretches outwards: it's heavy and it feels like it needs holding up with my arms. My balance has shifted forwards, and I feel a need to counter it by wearing a rucksack all the time. My navel is moving further and further from my spine, my uterus is pushing my lungs upwards, and the foetus is kicking my ribcage so hard I'm afraid it's going to break. I feel enormous, like a constipated whale, and I don't know how I'm going to be able to stand this for several more weeks. But I'm lucky that it's my body that surrounds my skeleton, not my skeleton surrounding my body. My tummy could keep growing outwards until I tipped over, but I would still be able to move about.

The emperor scorpion (*Pandinus imperator*) has been carrying her young for as long as me, and now she's about to give birth.[266] She can hold up to thirty-two young in her abdomen, but unlike me, her skeleton is on the outside. It takes the form of a series of hard plates along the top of her body, which protect her from danger. At the same time, it restricts the amount of space inside her. Now, at the end of her pregnancy, she looks as though someone has inflated her with a bicycle pump. The plates of her skeleton are as far apart as they can go. They no longer fully overlap, leaving her completely unprotected in the gaps where they would normally meet. She moves more slowly and can no longer travel as far as she could before. She hides in a dark crevice or behind a stone, saving her energy. In the event it enters anyone's head to approach her, she raises the venom-filled stinger at the end of her tail, so it hangs threateningly over her body, and waves her huge claws. It makes quite an impression: at up to twenty centimetres long, the emperor scorpion is one of the largest scorpion species in the world.

Even though they are members of the arthropod group, and therefore related to both insects and crustaceans, scorpions are not insects, like earwigs, and nor are they crustaceans, like lobsters. Instead, they are arachnids, like spiders and mites. They are extremely old – the oldest scorpion fossil ever found is more than 430 million years old[267] – and they look more or less the same now as they did then. We don't know when they came up out of the water, but it was likely during the Silurian or the Devonian period, or 430 or 300 million years ago. What we do know is that at some point after venturing onto land they started giving birth to live young. While the dinosaurs came into being, walked the earth, and went extinct, while mammals were gradually taking over the world, scorpions were hunting small creatures with their stingers, gripping their victims with their pincers, and giving birth to live young.

Thirty-seven weeks ago, our female met a male, and on this occasion they each decided not to eat the other. Perhaps she flaunted some pheromones that told him she was ready to accept his sperm. In any case, they held each other's claws and danced, as he pulled her over the packet of sperm he'd placed on the ground. She took up the sperm through her genital opening, and her eggs were fertilised.

Then came the long pregnancy, in which the foetuses slowly grew inside her system of ovaries, which consists of three long channels connected by four or five perpendicular tubes. Each foetus grows in a little bulge off one of these channels, gradually getting larger and larger as the mother's skeletal plates grow further and further apart from each other. They have neither umbilical cords, nor egg yolks to sustain them.[268] The foetuses' mouth parts develop first, and they bite onto the bulge they are in, getting food from the mother directly into their mouths.

The emperor scorpion's delivery takes longer than a day. She stretches some of her eight legs until she is standing high above the ground, using another set of limbs to pick up the young as they are born, one after the other, tail first. She throws them up onto her back so they don't fall on the ground, and they'll spend the next few weeks there until they shed their skins for the first time. Up to that point, they are soft and defenceless. Until this first skin-shedding, they won't eat anything, living instead off the energy reserves stored in their bodies from the food they received inside their mother. If she doesn't find enough food for herself, or eventually, for her young, she can end up eating some of them as a last resort – it's important for her to survive and have a new brood, so a juvenile or two can be sacrificed to keep her going.

Week 38

My uterus is full. It's getting more cramped for the foetus now, and he alternates between kicking against my ribcage and my bladder. He wants to stretch out his arms and legs and he has more strength to shove at my intestines as he tries to make more space for himself. Is it starting to feel uncomfortable for him? Does he want out? He's not even alone in there. He shares the space with an organ that's all his own, one he's been building ever since he was a little cluster of cells embedded in the lining of my uterus. The placenta is his life source, it gives him the nourishment he needs from me.

His big sister turns four: the apartment is full of balloons and nursery kids are running wild. They eat pizza and cake, and as their sugar intake rises, so does the volume. My foetus grows stiller and stiller the louder the other kids get. Is he listening to them? Is he learning how birthdays are celebrated, how other kids' voices sound?

When I give birth to him a few weeks from now, I will also deliver the placenta. He'll come first, still tethered by his umbilical cord. We'll cut it to mark the fact he's no longer attached to me, but the umbilical cord will still be attached to the placenta, of course, and it's the placenta that constitutes the real boundary between him and me. After all, it's the foetus who builds up the placenta, who has the recipe for making it, even if he does so with nutrients I give him. To begin with, when he's really tiny, it surrounds the foetus completely. But as he grows, parts of it are pushed back and it ends up to one side of him.

My blood never comes into contact with his. If it ever did, my immune system would treat him as an intruder and try to reject him. The placenta grows into the lining of my uterus, but it doesn't break through to the large, powerful muscles that form its outer wall. On the side facing the foetus, the placenta is smooth, and blood vessels branch out from the umbilical cord towards the mucous membrane lining of the uterus, resembling a tree diagram of how species have evolved from a common ancestor. On the side facing my uterine wall, the placenta is rough, consisting of a great many villi – little lumps in the outermost cell layer that make the surface area very large, just like in our intestines. The villi extend into the lining of my uterus, where the spiral arteries – the ones the foetus took control of in week three – discharge blood. In the area around the spiral arteries, nutrients and oxygen are taken up and transported out of my blood, across the cell layer in the villi, and into his waiting vessels.[269]

When I've delivered both him and his nutrient-hijacking organ, the midwife will examine the placenta and make sure it looks normal. If it does, she throws it in the contaminated waste bag, to be sent for incineration. The vast majority of animals eat their placenta; humans generally don't – we and whales, seals and camels tend to abstain, though our closest relatives, the other great apes, do eat theirs. The placenta is usually eaten either by the mother or by other females nearby, though in some rat and hamster species, others in the group, like a father or an older sibling, might be the ones to enjoy this nutritious, hormone-rich hunk of flesh.

There's little to indicate that humans before us ate their own placentas, even though dried placenta features in some traditional Chinese medicines, as a treatment for both men and women. It was used to treat chronic coughs and erectile dysfunction, but it wasn't something the birthing person would eat themselves, reserved for her after the birth as is the case for most other animals.

There are many reasons why certain animals eat the placenta. It can aid the removal of membranes and mucus covering the mouth and nose of the newborn, and strengthen the bond between mother and child. It is also highly nutritious, and in some species, speeds up the onset of milk production in the mother. Last but not least, disposing of the placenta by eating it can make it more difficult for predators to sniff out newborn young.[270]

When the human placenta is delivered, it is a blueish red disc, two to three centimetres thick and approximately fifteen to twenty centimetres in diameter, with a weight of five to six hundred grams.[271] Since the 1970s, interest in eating the placenta has grown in Western countries, probably owing to the rise in interest in natural remedies, and there are companies that will offer to turn it into a powder packaged in capsules for the birthing person to eat after the birth.

Few studies have been done on the effect on humans of eating the placenta, but those that do exist show that eating capsules of dried placenta provides none of the benefits the capsules are promoted with, such as increased levels of hormones, increased milk production, or faster weight gain in the newborn. Some people swear by eating the placenta fried like a steak, or in a smoothie, but there's much debate over whether this has any positive effect. Those in favour assert that it reduces the risk of postpartum depression, but eating it raw or inappropriately prepared can lead to food poisoning, and the placenta may even contain dangerous levels of hormones or other substances.[272][273]

Of all the organs mammals have evolved, the placenta exhibits the greatest variation.[274] We mammals can be very different from each other – from the bat to the human to the dolphin to the giraffe, but our placentas are even more distinct. And placenta-like structures are present not only in mammals. They've evolved separately many times and are also seen in reptiles, amphibians and fish. In addition, it's not only vertebrates that can have placentas: similar structures are also found in velvet worms and some arthropods (the group that incorporates insects, arachnids and crustaceans, among others),[275] animals that are as different from us as it's possible to imagine.

Mammalian placentas can go from being barely in contact

with the mother's tissue, asking politely for a few nutrients, to invading the tissue and taking control of her bloodstream, as in humans.[276] The placenta is an important part of the struggle for resources between the pregnant animal and its foetus, between me and the little one who'll soon emerge, and it is most invasive in the great apes. This may be because our brains are so large. The brain of a primate is proportionately five to ten times larger than that of the average mammal.[277] It takes a lot of resources to build a foetus's brain, and there's a lot that can go wrong if it's not assembled correctly. When the foetus takes control of the spiral arteries and sees to it that it gets enough nutrients, it's safeguarding its own brain. The exchange of nutrients through the placenta is intensive. So intensive that some of the foetus's cells will enter the pregnant person's body. They can stay there a long time, perhaps for the rest of their life.[278]

The nausea returns when I think about people eating their placentas, searing them like a steak, or blending them into a smoothie. I don't want to think about the weird blood cake my baby has created. He's the one I want, not the extra organ he needs to have in order to draw nourishment from my body. I want to cut his umbilical cord, to feel his skin against mine, not deal with the placenta, the organ that has enabled him to grow so large, to get so many resources from me, and in the end, to become a human being.

Week 39

Inside a rotten tree stump somewhere in the tropics, a pregnant velvet worm (Onychophora) has made herself at home. She might not be that big – the largest species are only fifteen centimetres long, and many are much shorter, just five millimetres from the head to the tip of the tail. A snake made of velvet, with soft-looking skin that water runs right off. A snake with legs – between thirteen and forty-three pairs that stick out from either side of her, with thin little claws at the ends. On the top of her head, she has two antennae and two little eyes; on the underside of her body, strong jaws she uses to crush insects and small spiders. And in her belly there are lots of squirming velvet worm foetuses that have been growing and growing and will soon come out.

Velvet worms aren't snakes, but nor are they insects, worms or crustaceans. They're in a group of their own and they have lived virtually unchanged on this planet for more than 500 million years. We've found about two hundred species, and their distribution around the world fits with the shape of the supercontinent Gondwana before it started breaking up a little more than 200 million years ago. They're distributed throughout tropical and subtropical regions that now lie far apart from one another, but were connected before the continents assumed their current form.[279]

The various species of velvet worm resemble each other from the outside: they look soft, have multiple pairs of legs and fairly similar bodies. But if you look at how they reproduce, they're very different. Some lay eggs, some produce eggs and keep them inside their bodies, some become pregnant and provide their foetuses with nourishment inside their bodies, just as we do. The species that have been studied produce between one and twenty-three young a year, and some species can be pregnant for up to fifteen months.[280] Some have foetuses at varying stages of development inside them simultaneously: some who will soon be born, and some younger ones who will need to stay where they are for a long time before they are ready to enter the world. Some store sperm for many years, and some may have organs that destroy sperm they don't want – sperm forced upon them by eager males. But that's not all. Some receive the sperm from males through their genital openings, delivered by a little growth on the male's head. And others don't use the genital opening, getting instead a packet of sperm deposited on their skin, where it melts into their bodies.[281]

Some males put their sperm, packed like a little gift called a spermatophore, on their head, before approaching a female. Do the females find this sperm crown a fetching adornment? Are they wondering whether they'd like to mate with the males based on how well they balance the sperm on their head? Are they charmed? Can they feel it in their claws? We don't know. Not many instances of mating have been observed, but in one of the few that have been described, the female held the male's head and his sperm crown against her genital opening throughout the act, which lasted at least fifteen minutes. We don't know why males make a sperm crown, or how they get the sperm up onto their heads, but perhaps it's an adaptation to the long, cramped hollows in rotting wood they call home, where there's not much space for acrobatics – perhaps head to bottom is the best space-saving method of doing it if your body is long and thin.

Maybe this is also the reason the females can absorb the spermatophores through their skin. In one of the most widely studied species, the male can place his sperm anywhere on the female's skin. Her body's blood cells race to the spot and make holes in both her skin and the sperm packet. After that, the sperm swim in through the blood-filled hollows of her body until they find the ovaries and get to work trying to fertilise the eggs.[282]

Somewhere or other out there right now, there's a pregnant, velvety sausage with a body full of velvety foetuses. Maybe she'll give birth soon, maybe she'll be pregnant for another six months. Maybe she barely notices that she's pregnant as she crawls around catching insects and tiny spiders, maybe her body feels heavy and not as cooperative as usual. Maybe she'll come across a male with a sperm crown and think he looks pretty silly, because she's not in the mood to mate again, or maybe she'll be charmed by him anyway. We know very little about velvet worms and how they reproduce – and even less about what they think about their situation.

Week 40

I'm having a check-up, and I let the midwife sweep my cervix with her fingers. This is intended to set labour in motion, which suits me because I'm keen to get this kid out – I'm sick of being pregnant. I've taken off my maternity trousers with their soft panel of stretchy fabric that covers my bump, and the big knickers with extra space at the front. I've lain down on a bed and put my feet up. Lying on my back is difficult with this big belly – I'm given a pillow to support me, but the foetus is pressing on my lungs and my breathing is a little laboured. The midwife sticks her gloved hand inside me and sweeps it over the door that's keeping the baby inside – pain shoots through my body. When she draws her hand out again, there's blood on the glove. My cervix is ripe, she says, trying to console me with the promise that something will happen soon – she can tell how fed up I am. I picture my cervix as an unripe banana that's hard to peel, and when you try, the flesh inside just turns to mush. After the check-up I waddle down to the bus stop and manage to find a seat on the bus without having to ask anyone to move. I don't know it yet, but I'll be pushing a pram the next time I take this bus home from the hospital.

Somewhere in the not so deep blue sea, a sand tiger shark (Carcharias taurus) is also getting ready to give birth.[283] This two- to two-and-a-half-metre-long grey shark has a streamlined body, a white belly, small eyes and sharp teeth. Sand tiger sharks live in the oceans off the coasts of Japan, Australia, South Africa and eastern North and South America. They get their name from their preference for shallow sandy beaches, but they range to depths of a hundred and ninety metres. They eat fish, crabs, squid, skates and other sharks that they hunt on the seabed,[284] where they patrol and reproduce, feeding their foetuses in a way few others do.

She's heavy now after forty weeks of pregnancy. Her young have grown inside her, and now there's only one left in each of her two uterine chambers. The rest of the fertilised eggs she started off with have been eaten by the first foetus in each chamber to hatch. When the two foetuses have eaten all the fertilised eggs, they go on eating unfertilised eggs until they're ready to be born. Instead of getting nutrients directly from her body via an umbilical cord, they eat their potential siblings.[285] Now the two shark foetuses are each a metre long, almost half the length of their mother. They'll soon be free, ready to fend for themselves from day one, swimming off immediately after birth. They've been well fed inside the uterus, and when they're finally released into the ocean, they'll have few natural predators. Perhaps it's worth sacrificing a few siblings to make young that are large enough not to get eaten as newborns.

The sand tiger shark's single functioning ovary can weigh up to eight and a half kilos and contain more than twenty-two thousand eggs. That's food enough for the growing foetuses, the ones big enough to devour the others.[286]

*

Another species that will soon give birth is a subspecies of the European fire salamander (*Salamandra salamandra bernardezi*), which, like many of its close relatives, is black with yellow markings – a clear warning that it's poisonous. It doesn't lay eggs like many other salamanders do, choosing not to risk it all in a pond that might dry out or be visited by hungry predators. It keeps its eggs inside its body and gives birth to live young after nine to twelve months, depending on temperature and other environmental factors. Whereas salamander eggs are generally laid in water, where they hatch and turn into larvae that then metamorphose into adults, European fire salamander eggs hatch inside the mother. Her young undergo their entire larval stage within her body, and only once they've passed through their gill phase and have fully functional lungs, does she push them out – they're born like mini adults, ready for life both on land and in water.[287]

When I give birth and have finally put the first, fundamental part of reproductive labour behind me, there will be others still carrying their foetuses for a while to come. Coelacanths (*Coelacanthiformes*) are a group of fish that were long considered extinct because they had only ever been seen as fossils. But one day, they were suddenly found to be swimming around in the sea, as they had been doing for millions of years, completely unbeknownst to us. Coelacanths are thought to be pregnant for around a year,[288] just like the bottlenose dolphins who talk to their unborn calves. It's still twelve weeks until they will give birth.

The Eastern grey kangaroo's baby crawled into her pouch back in week five, but the mother's reproductive labour is still not over. When I give birth, she'll have the baby in her pouch for another thirteen weeks before it finally leaves her body.

The sperm whale (*Physeter macrocephalus*) is pregnant for fourteen to sixteen months and has calves at five-to-six-year intervals. They nurse their calves for years.[289]

The European lobster (*Homarus gammarus*) carries her eggs for two years. First she builds up her roe internally for a year, before releasing it to be fertilised by sperm she has been storing and attaching the eggs to her pleopods, or swimming legs, where they remain for a year before hatching.[290]

The African elephant (*Loxodonta* spp.) is pregnant for twenty-two months – almost two years. When I give birth, the African elephant will still have thirteen months of her long pregnancy to go, which is a record among mammals alive today,[291] but it's not she who is pregnant the longest.

That distinction belongs to the frilled shark (*Chlamydoselachus anguineus*), which lives at depths of up to thirteen hundred metres in many locations around the world. This one-and-a-half-metre-long, eel-like shark, which is thought to be a kind of ancient living fossil because of how little it has changed over millions of years, can be pregnant for up to three and a half years before giving birth to between two and ten young.[292]

Ever since life first began, 3.5 billion years ago, we've been reproducing. Now it's my turn. This evening my uterus will begin to contract rhythmically; I'll sense that these are not just practice contractions, but full labour pains, and I'll make my way to the hospital, back to the midwife who swished her fingers over my cervix. When she meets me at the door, she says she knew I'd come that evening, and that gives me the confidence to let my body take over. That very night my uterus pushes out a healthy, adorable little baby.

Right after I've delivered the baby, I deliver the placenta, and we're no longer physically linked, my child and I. With my newborn at my breast, I suddenly become intensely interested in the placenta – now it's outside my body, it no longer induces nausea. Perhaps it's all the hormones and the joy of having the birth behind me, or perhaps it's some primeval instinct that makes me want to be close to it, to smell it. The midwife shows me the contents of the metal dish, pointing out the blood vessels that branch out from the umbilical cord across the red disk; she calls it the tree of life. She checks it for blood clots and discolouration, but it's perfect. As I waddle down the postnatal ward, using the transparent plastic trolley my baby is lying in as a kind of walker, the midwife drops the once life-giving organ into the contaminated waste bag. It has done its job.

We've got something in common, the frilled shark, the sand tiger shark, the fire salamander and I: we carry our young inside us. But this similarity is not based on a common evolution. Giving birth to live young has evolved multiple times, with distinct origins. The way we provide our foetuses with nourishment is also different. My foetus built a placenta and burrowed its way into the lining of my uterus, taking control of my bloodstream – I fed him directly from my body. The sand tiger shark feeds its foetuses on its eggs, while the frilled shark's foetuses live off their yolks, which she made as the eggs were maturing inside her, before the foetus was enclosed in its foetal membrane. We don't know whether the fire salamander nourishes her young while pregnant, or if they have to make do with their egg yolks, but we do know that the ability to birth live young has developed many, many times across the tree of life – it's not a new invention.

Scorpions started doing it at some point after they came up on land between 430 and 300 million years ago. Some species had umbilical cords and gave birth to live young even earlier. A 380-million-year-old placoderm, an extinct kind of fish, was fossilised with an embryo inside her uterus, attached to her by an umbilical cord.[293] The first mammals existed around 180 million years ago, and at least 140 million years ago they split into two groups: marsupials and placental mammals.[294] It's hard to ascertain exactly when different events took place in evolutionary history, and our understanding can shift as new discoveries are made and new technologies develop. But we know our own, very special placenta developed some time before the two groups split off from each other, and since then, the extremely intimate nutrient exchange we have with our foetuses has evolved a great deal in placental mammals like us. Many reptiles give birth to live young, but they evolved that characteristic later than we mammals did.[295] The ichthyosaurs – dolphin-like marine reptiles that lived from approximately 250 to 100 million years ago – are an example of a group that gave birth to live young before we mammals came up with it, but we don't know if they nourished their foetuses with resources from their own bodies, or if the next generation lived off the yolks in their eggs inside their mother.[296]

But why do we do it? Why do we carry our foetuses around with us, why do any of us give them nutrients from our own bodies? Wouldn't it be easier to simply lay an egg, to avoid one's body being taken over by a parasite-like organism?

When I carry my foetus instead of laying an egg in a nest, sticking it to the underside of a leaf, or burying it in the sand, I'm protecting it from drought, floods, or predators on the lookout for a nutritious meal. By keeping it inside me, I can produce relatively small eggs, and then increase my investment when I know they've been fertilised and are therefore more likely to amount to something. Those species who lay eggs, releasing them into a body of water, or guarding them in a nest, have to give the growing foetuses all the nutrients they're going to need from the outset. I don't have to prepare a full packed lunch beforehand, before I know for sure there's going to be any point in doing so. Neither do I have to take care of an egg that's outside my body and can't be moved. Pregnant humans don't have to sit still for weeks or months – apart from those of us who are overpowered by our foetuses, betrayed by our own body's urge to vomit.

On the other hand, you can turn all these arguments on their head. Egg-laying animals can reproduce more often and have more children – they don't tie up their bodies in reproductive labour for long periods. An egg-laying female doesn't get big and slow, and she can abandon her eggs to save herself from being eaten should a predator find her.

Giving birth to live young has evolved more than a hundred and fifty times, in different animal groups, so it obviously has something going for it.[297] It's hard work for the pregnant animal, but it does the job: it produces young, even if the process is convoluted.

Like all other offspring produced through sexual reproduction, my foetus began life as a little unfertilised egg. The cold water coral released its eggs into the water, while I kept mine inside; the coral's egg found sperm in the water, while mine travelled from the uterine tube to my uterus as it decided which sperm to allow in. My egg fought its way into the lining of my uterus, the eider's egg was laid in a snug nest outside her body, the Suriname toad's were stuck to her back. I had to carry my foetus inside me throughout, while the kangaroo gave birth early and let her pouch do the rest. While my belly grew outwards, away from my spine, the emperor scorpion's skeletal plates stretched apart from each other, and while my baby took nutrients from the placenta, from me, the sand tiger shark gobbled up its siblings, the foetuses of the Pacific beetle cockroach got uterine milk, and the kiwi chick lived off its giant egg yolk.

I can't lay an egg: my ancestors have bound me to my uterus. I'm tired now, just as the eider duck was when her eggs finally hatched, as the Atlantic wolffish was after lying still with no food for seven months. In the end it goes well for me – my baby comes out, we survive, both of us, and all I need are two stitches to my torn labia. For the time being, at least, he's the last in a long line of organisms who have reproduced, from the first cells dividing in the primordial soup to the enormous variation in methods of passing on genes that we see today.

The most successful reproductive strategy is the one that produces as many able offspring as possible without the parents expending all the resources they have available. The sea anemone, the spotted hyena, the Namaqua chameleon, the Damaraland mole-rat, the little Japanese aphid, me, and all the other species that exist in the world right now are examples of that.

During pregnancy it might feel like a consolation that I don't have to give birth through a pseudo-penis or deliver an infant with a body weight of up to 45 per cent of my own. And even if the breastfeeding and the wakeful nights that are to come are exhausting, I'm not about to die, like the giant Pacific octopus, or be eaten by my own children, like the African social spider. Evolution has given me nausea, vomiting, swollen legs, pelvic pain and a heavy belly, but also cooperation, the opportunity to see my young grow up, and grandparents who lend a hand. It's worked out alright, the long road from the primordial soup to the humans we are today. I may have been cursing my ovaries as I hunched over the toilet bowl to vomit, but the knowledge that I was doing so to protect my foetus, even if my body was freaking out a bit, offered a kind of comfort.

Through the long, heavy days of a human pregnancy, it gave me a little respite to know that most of the things happening in my body were happening for a reason – not because of some supernatural power of destiny, but because through the ages they have made sense from an evolutionary perspective. And when the baby emerges, it's generally worth it.

Week 0

After two days on the postnatal ward, filled with breastfeeding guidance, blood tests and plastic trays of hospital food in bed, I pack up my pyjamas and my partner pushes the pram into the lift. The huge maternity pad chafes against my sore skin, tugging at the stitches holding my torn vulva together. Luckily, we haven't got far to go. We take the lift three floors down to the ground floor, walk the few metres to the bus stop, then board the bus. As it snakes its way through the city, a squeaking noise emerges from the pram – he wants food, and he wants it now. I pick him up, and the squawking is replaced by a powerful latch around my nipple, as an older woman leans across the gangway to congratulate us. She reaches into my private space and strokes the back of his hand gently as he nurses – this newborn human is already getting attention, and even though he's not part of this old lady's tribe, she simply has to greet him, just like female Hanuman langurs and elephants flock around newborn baby Hanuman langurs and elephants. There on the bus, we're suddenly a little community for the twenty minutes our journey takes. She tells me she breastfed three kids and that we're doing so well, everything's going to be fine.

I drag myself up the stairs all the way to the third floor, while my partner juggles the baby and the bag with all our things, then I carefully arrange myself on the sofa between cushions and blankets. My pelvis is aching, my insides are behaving strangely now there's suddenly so much space, and it feels like I need to keep my hand between my legs to stop my uterus falling out. My partner unpacks our new little person from his muffler and woollen blankets, his woolly jumper and hood, and his chunky trousers. The little one flings his arms about and starts crying the moment he's lifted out of his warm nest, but quickly calms when I put him on my body. We lie chest to chest: he inhales the scent of my neck, and I inhale the scent of his hair.

We've sent the four-year-old on an Easter holiday with her grandparents and cousins. We're entering a new era of engorged breasts, dirty nappies and sleeping in shifts. While my body gradually stops bleeding from the wound the placenta left in my uterus, while my uterus is drawing back in on itself, stimulated by the hormones released in my body when I breastfeed, we're establishing a new symbiosis. He may not be controlling my blood flow anymore, but his hungry little snorting noises control my body, and my nipples start leaking milk at the mere suggestion he might be hungry. I've got my body back – we're two separate individuals now – but he wants me all the time. My body follows his command, and my partner has to follow mine – making sandwiches, changing nappies, rocking the baby and letting him lie skin-to-skin so he feels the comfort of a heartbeat while I take a shower.

The kangaroo gave birth long before me, but her baby still hasn't begun to move from the pouch. The reindeer calf can manage on its own in the herd from the age of one, whereas the titi monkey infant clings to its dad and stays with its parents for several years. Baby chimpanzees are breastfed until they are around five, but they have to stay close to their mother for the protection and knowledge she offers until they turn ten. Orcas need their mothers when they themselves become mothers. The baby is out in the world, but the reproductive labour will continue for many years – it just takes another form.

Evolutionary Timeline

This is an overview of some of the evolutionary events mentioned in this book. It's hard to confirm precise dates, because evolution happens slowly and because we are constantly getting new data and new models, and with them, new knowledge. All these are rough estimates.

- Now: My baby is born.[298]
- 200,000 years ago: Earliest modern humans (*Homo sapiens*).[299]
- 40–50m years ago: Earliest instances of menstruation.[300]
- 60m years ago: Earliest primates.
- 65m years ago: Cretaceous–Palaeogene extinction event, non-bird dinosaurs become extinct.
- 130m years ago: Earliest flowering plant species.
- 140m years ago: Mammals split into two branches: placental mammals and marsupials.
- 180m years ago: Ancestors of the monotremes branch off from the rest of the group that will in

time evolve into the other mammals. Monotremes, such as the platypus, lay eggs and have milk glands but no teats.
- 230m years ago: Earliest dinosaurs.
- By 318m years ago: Earliest amniotic egg, which has a membrane to prevent it drying out, in a common ancestor of reptiles, birds and mammals.[301]
- 375m years ago: Earliest tetrapods, fish-like creatures with strong skeletal structures allowing them to "walk" in shallow water.
- 850m years ago: Earliest multicellular organisms.
- 2,000m years ago: Earliest sexual reproduction.[302]
- By 3,500m years ago: Earliest life, known as the primordial soup.

Glossary

All definitions of words are based on definitions provided in the Oxford English Dictionary (oed.com), unless otherwise stated.

Adaptation
A process of change or modification by which an organism or species becomes better suited to its environment or ecological niche, or a part of an organism to its biological function, either through phenotypic change in an individual or through an evolutionary process effecting change through successive generations.

Amniote
Any vertebrate whose embryo develops within an amnion and chorion (membranes surrounding the foetus). Includes reptiles, birds and mammals.

Anthozoa
A large class of marine invertebrates comprising forms which

are predominantly sessile and lack a medusoid stage in the life cycle, including sea anemones, sea pens and most kinds of coral. Many species' larval stages are dispersed freely in water.

Cloaca
In birds, reptiles, most fish and the monotreme mammals (egg-laying mammals including platypuses): the common chamber or cavity into which the digestive, urinary and reproductive tracts discharge their contents.

Crustaceans
A large group of mainly aquatic arthropods comprising crabs, lobsters, shrimps, woodlice, barnacles and many minute forms, which are very diverse in appearance but generally have two pairs of antennae, four or more pairs of limbs, and several other appendages.

Embryo
The unborn, unhatched or incompletely developed offspring of an animal. The term is now most narrowly applied to the human organism from the point, usually in the second week after fertilisation and just prior to implantation, when its cells become differentiated from those of the trophoblast, until the end of the eighth week, when the organs begin to develop and it is termed a foetus. The term embryo is sometimes extended (esp. in popular and non-technical use) to include the zygote or fertilised ovum before cell differentiation, or applied to the foetus during later development.

Embryophagy
The act of one embryo cannibalising another for food in utero; at present only characterised in some species of sharks.[303]

Embryonic diapause
Embryonic diapause is a reproductive strategy in which the embryo's development is arrested in the blastocyst stage, i.e. when the embryo is a small ball of cells. The presumed function is to make it possible for the species to uncouple mating from birth, in order to ensure that both occur at the time that is most advantageous for the species, as for example with the brown bear, who can mate early in the season and use the autumn to fatten herself up, rather than searching for a partner. The diapause can be facultative: initiated by physiological conditions in the body of the gestating animal; or obligate: present in all pregnancies in that species.[304]

Evolution
The transformation of animals, plants, and other living organisms into different forms by the accumulation of changes over successive generations; the transmutation of species; the origination or transformation of an organism, organ, physiological process, biological molecule, etc., by such a series of changes.

Foetus
The developing offspring of a human or other viviparous animal in the period after the major structures of the body have been formed.

Genes and gene variants

The basic unit of heredity in living organisms, passed down from generation to generation. All the individuals in a species share the same set of genes, but have differing alleles. However, people commonly talk about genes when they really mean alleles, and there are a number of occasions in this book where I use the term gene instead of allele in order to simplify the language.

Hemipenis

A hemipenis is one of a pair of reproductive organs of male squamates (snakes, lizards and worm lizards) that are usually held inverted within the body until mating occurs. They come in a variety of shapes and may have spikes or hooks. Hemi means "half". One testicle is connected to each hemipenis.[305]

Hermaphrodite, sequential and simultaneous

A hermaphrodite is a sexually reproducing organism that produces both male and female gametes. We distinguish between sequential hermaphrodites, in which the individual first develops as one sex, but can later change into the opposite sex, and simultaneous hermaphrodites, in which an individual possesses fully functional male and female genitalia.[306]

Isopods

Isopods are an order of crustaceans. They can be found in marine and freshwater environments, and on land. Woodlice, which are discussed in this book, are a sub-order of isopods.

Iteroparity
Iteroparous species are species that have the ability to reproduce several times in their life. Examples include humans and eider ducks.[307]

Marsupials
Mammals of the order Marsupialia, characterised by the bearing of very immature young which are typically nursed in an abdominal pouch, whose extant members include kangaroos, opossums, etc.

Matriphagy
Matriphagy is the consumption of the mother by her offspring, as in the African social spider.[308]

Nymph
The larva of insects that undergo incomplete metamorphosis, frequently resembling the adult in form though sexually immature. These species of insect do not go through the pupal stage.

Organism
An individual animal, plant, fungus or single-celled life form.

Parasite
An organism that lives on, in, or with an organism of another species, obtaining food, shelter, or other benefit. The parasite obtains nutrients at the expense of the host organism, which it may directly or indirectly harm.

Parthenogenesis
Reproduction from a gamete without fertilisation, occurring most commonly in invertebrates and lower plants.

Placental mammals
Mammals characterised by the development of a true placenta in the female, as distinct from monotreme or marsupial mammals.

Primates
An order of mammals that includes humans, apes, monkeys and prosimians, which in general have relatively rounded skulls and flattened faces, short jaws and noses, forward-facing eyes and an opposable digit on each foot (except in humans), and which are typically agile tree-dwellers.

Primordial soup
A solution, rich in organic compounds, which is thought (by many scientists) to have been the environment in which complex biological molecules and hence life originated.

Reproduction
The production by living organisms of new individuals or offspring; the perpetuation of a species by this process; the power of reproducing in this way. Also: the process or mechanism by which this takes place, whether sexual or asexual in nature.

Reptiles
Animals of the vertebrate class Reptilia, the members of which are characterised by their dry impervious skin covered in horny scales. The class includes snakes, lizards, turtles and crocodiles.

Semelparity
Semelparous species are those that always die after their first reproductive event, like the giant Pacific octopus.[309]

Sexual selection
The evolutionary theory, originally proposed by Darwin, of the preferential reproduction of male organisms with characteristics that favour their success in competition with other males, either directly or through mate choice by females, intended to account for the development of features such as large size, elaborate horns, ornamental coloration, etc.

Squamata
The largest order of reptiles, comprising lizards and snakes. It does not include reptiles such as turtles, crocodiles and tuatara (the last surviving member of the once diverse group Rhynchocephalia).[310]

Symphysis
The union of two bones or skeletal elements originally separate, either by fusion of the bony substance or by intervening cartilage: used esp. of such union of two similar bones on opposite sides of the body in the median line, as that of the pubic bones (symphysis pubis).

Vertebrate

Animal belonging to the subphylum Vertebrata, i.e., one having a backbone or spinal column. Includes fish, amphibians, reptiles, birds and mammals.

Afterword

Biology is variation, variation in most traits of most organisms. Much of what I have written here uses generalisations – that humans are pregnant for forty weeks, for example. We tend to say that a human pregnancy is forty weeks long, but in reality it isn't. It's standard practice to count the length of a pregnancy from the first day of the last menstrual period, but fertilisation of the egg does not occur until around fourteen days after that point. In Norway, as in many other countries, we still consider a pregnancy to be forty weeks long, because the first day of the last period is the clearest indicator we ourselves have of when we became pregnant. This method assumes that ovulation – the point in time when we can actually become pregnant, occurs exactly fourteen days after the first day of the individual's period, but there can be significant variation in this. The length of human pregnancies can vary so much that, in Norway, all deliveries that take place between thirty-seven and forty-two weeks are considered to be full-term deliveries.[311]

*

I have chosen to base my comparisons of humans with other animals on the time during which the mother is deemed to be pregnant – that is, from the first day of the last menstrual period, not from when the egg was fertilised, even though that is the point from which we start calculating other animals' gestations. Technically, I should perhaps have calculated human pregnancy from week two, in order to be able to make comparisons with the lengths of time taken by other animals to get from the point of fertilisation to the point at which they complete the reproductive event, but since my focus is on the parent, and not on the fertilised egg, I've chosen this approach. Moreover, the first day of the last period is the point at which our bodies begin their preparations for pregnancy – the spontaneous release of an egg is reliant on the expulsion of the uterine lining and the start of the hormonal cycle, whereas many other species' pregnancies begin with copulation, which is required for an egg to be released. In yet other species, as in many birds, the reproductive period is thought to include the work of the parents from the time the eggs are laid to when they hatch, while the enormous work of creating the actual egg – that of the kiwi, for instance – has historically not been counted.

In this book I describe biological sex and refer to individuals as "females" and "males" and "mothers" and "fathers". Females produce eggs, while males produce sperm. This doesn't necessarily have any bearing on how reproductive labour is distributed across the sexes, as I've demonstrated with a large number of examples throughout the book. When I call animals – including humans – males and females, mothers and fathers, this is based on the reproductive organs that have historically been visible, not on genes, hormones, how different individuals express themselves, or how they feel. We humans have a range of possible expressions of sexual and gender identities and roles that cannot be viewed in terms of other species' ways of being or dividing up their reproductive labour. The use of the terms mother and father to refer to humans and other animals in this book must not, in other words, be seen to subscribe to biological determinism – which asserts that humans should behave or think in a certain way on the basis of which external sex organs they had at birth.

Referring to other animals as "mother" and "father" is anthropomorphism – giving animals human characteristics – and can easily lead to us applying our own gender roles and assumptions about what it is to live on this planet to other species. This is problematic, but it can simultaneously increase our understanding of the fact that other animals besides humans have needs and feelings, and it is with this intention that I have used these terms.

The events described in the first person in this book are based on my own two pregnancies. Everything the first-person narrator experiences are things that can happen during human pregnancy, but they didn't necessarily take place in the exact manner or order described here.

Thanks

The idea for this book came when I was pregnant with my first child, lying in a dark room and vomiting for weeks, while the world outside spun past me. It didn't make sense that pregnancy could make me so sick – biology must offer a better way to reproduce, right? Turns out that for us humans, it doesn't, and yet I still got pregnant again. Looking back, the products of these two pregnancies, Ea and Bjørn, were totally worth it. Thanks for making my life so exciting and beautiful! (Though it would be nice if you could start waking up a little later in the morning.)

Thanks to the Norwegian Non-Fiction Writers and Translators Association for understanding that reproductive labour sometimes takes priority over the work of writing. Thanks to my editor Finn Totland for believing in this project and demonstrating patience on a pandemic scale. Thanks to Fredrik for giving me time off work. Thanks to all the scientists who have patiently answered my questions and given me access to scientific articles. Without your research and the time you took to answer my sometimes rather basic enquiries, there would have been no book. Any mistakes or misunderstandings of the scientific material found here are my responsibility entirely. Thanks to Oda Noven, Trude Myhre, Ane Sydnes Egeland and Ingerid Salvesen for comments on earlier drafts. A special thanks goes to Ellen Støkken Dahl for reading me so well, and to Kjetil Lysne Voje for nudging my use of scientific terminology in the right direction – any mistakes, simplifications or inaccuracies are down to me.

Finally, thanks to evolution for brains, our capacity for cooperation, and grandparents.

Notes

1. This comparison between a foetus and a parasite is described in Bainbridge, D. (2003). *Making Babies: The Science of Pregnancy.* Harvard University Press.
2. Emera, D., Romero, R. & Wagner, G. (2012). The Evolution of Menstruation: A New Model for Genetic Assimilation. *BioEssays,* 34(1).
3. Moen, F.E. & Svensen, E. (2008). *Dyreliv i havet,* (5. utg). Kom forlag. (In Norwegian. Alternative source in English: The Marine Life Information Network (n.d.) Plumose anemone (Metridium senile). Accessed 11/04/2024 https://www.marlin.ac.uk/species/detail/1185).
4. The clonal plumose anemone is fortunate enough to be able to reproduce both sexually and asexually.
5. Pickrell, J. (2019, 23.10). *How the earliest mammals thrived alongside dinosaurs.* Nature news feature. https://www.nature.com/articles/d41586-019-03170-7
6. The number of bacteria after 40 weeks would be approximately 2^{20160}, although growth will always be limited by access to space and nutrients – as with us humans.
7. Havforskningsinstituttet (2020, 23.06). *Norske korallrev.* https://www.hi.no/hi/temasider/hav-og-kyst/norske-korallrev (in

Norwegian. Alternative source in English: The Marine Life Information Network (n.d.) A cold water coral (Lophelia pertusa). Retrieved ed 11/04/2024 from https://www.marlin.ac.uk/species/detail/1806).

8 Perth Cichlid Society (2016, 3.08). *Fish of the Month – Ctenochromis Horei*. http://www.perthcichlid.com.au/forum/index.php?showtopic=63151

9 Zimmermann, H., Blažek, R., Polačik, M. & Reichard, M. (2022). Individual experience as a key to success for the cuckoo catfish brood parasitism. *Nature communications*, 13(1723).

10 Emera, D., Romero, R. & Wagner, G. (2012). The evolution of menstruation: A new model for genetic assimilation. *BioEssays*, 34(1).

11 Brochmann, N. & Dahl, E.S. (2018). *The Wonder Down Under*. Yellow Kite. (Tr. Lucy Moffatt) (Originally published 2017 as *Gleden med skjeden*. Aschehoug.)

12 Emera, D., Romero, R.& Wagner, G. (2012). The evolution of menstruation: A new model for genetic assimilation. *BioEssays*, 34(1).

13 Cohen, M., Hawkins, M.B., Stock, D.W. & Cruz, A. (2019). Early life-history features associated with brood parasitism in the cuckoo catfish, Synodontis multipunctatus (Siluriformes: Mochokidae). *Phil. Trans. R. Soc. B*, 374.

14 Eckbo, N. (2021). *Keiserpingvin* i Store norske leksikon. https://snl.no/keiserpingvin (in Norwegian. Alternative source in English: Rafferty, J. P (2024) Emperor penguin in *Britannica* https://www.britannica.com/animal/emperor-penguin).

15 Wikipedia. (n.d.). *Emperor penguin – Courtship and breeding*. Accessed 17/04/22 at https://en.wikipedia.org/wiki/Emperor_penguin#Courtship_and_breeding

16 Pinshow, B. & Welch, W.R. (1980). Winter Breeding in Emperor Penguins: A Consequence of the Summer Heat? *The Condor*, 82(2).

17 Krause, W.J. & Krause, W.A. (2006). *The Opossum: Its Amazing Story*. University of Missouri Columbia.
18 Byrne, M., Hart, M.W., Cerra, A. & Cisternas, A. (2003). Reproduction and Larval Morphology of Broadcasting and Viviparous Species in the *Cryptasterina* Species Complex. *Biol. Bull.*, 205.
19 Byrne, M. (2005). Viviparity in the Sea Star *Cryptasterina hystera* (Asterinidae): Conserved and Modified Features in Reproduction and Development. *Biol. Bull.*, 208.
20 University of California, Davis. (2012, 24.07) *Superfast evolution in sea stars*. ScienceDaily. www.sciencedaily.com/releases/2012/07/120724104638.htm
21 De Waal, F.B.M. (2006, 01.06). *Bonobo sex and society.* Scientific American.https://www.scientificamerican.com/article/bonobo-sex-and-society-2006-06/
22 Hamzelou, J. (2022, 10.01). *What dolphins reveal about the evolution of the clitoris.* New Scientist. https://www.newscientist.com/article/2303662-what-dolphins-reveal-about-the-evolution-of-the-clitoris/
23 Rukke, B.A. (2021, 18.03). *Veggedyr*. Folkehelseinstituttet. https://www.fhi.no/nettpub/skadedyrveilederen/veggedyr-og-andre-teger/veggedyr/ (In Norwegian. Alternative source in English: Centers for Disease Control and Prevention (2017) Bed Bugs. Accessed 11/04/2024 https://www.cdc.gov/dpdx/bedbugs/index.html)
24 Elven, H. & Aarvik L. (2022, 30.03). *Tovinger Diptera.* Naturhistorisk museum, Universitetet i Oslo / Artsdatabanken. https://www.artsdatabanken.no/Pages/135156/Tovinger (In Norwegian)
25 Attenborough, D. (2005). *Life in the Undergrowth*. [Nature documentary]. BBC
26 Morrow, E.H. & Arnqvist, G. (2003). Costly Traumatic

Insemination and a Female Counter-adaptation in Bed Bugs. *Proc. R. Soc. Lond. B*, 270.

27 Nesheim, B.-I. (2022). *Graviditet* i Store medisinske leksikon. https://sml.snl.no/graviditet (In Norwegian. Alternative source in English: Orlowski, M. & Sarao, M.S. (2023, 01.05) Physiology, Follicle Stimulating Hormone. *StatPearls*. https://www.ncbi.nlm.nih.gov/books/NBK535442/).

28 Nesheim, B.-I. (2022). *Eggløsning* i Store medisinske leksikon. https://sml.snl.no/eggl%C3%B8sning (In Norwegian. Alternative source in English: Huffman, J.W. (2024, 24.04). *Pregnancy*. *Encyclopedia Britannica*. https://www.britannica.com/science/pregnancy).

29 Pietsch, T.W. (2005). Dimorphism, Parasitism, and Sex Revisited: Modes of Reproduction Among Deep-sea Ceratioid Anglerfishes (Teleostei: Lophiiformes). *Ichthyological Research*, 52.

30 Of course, there are some exceptions – such as bees that produce haploid drones when the eggs go unfertilised. See e.g. Britannica, T. Editors of Encyclopaedia. (2018, 23.01). *Chromosome Number*. *Encyclopedia Britannica*. https://www.britannica.com/science/chromosome-number

31 Britannica, T. Editors of Encyclopaedia (2022, 9.09). *Parthenogenesis*. *Encyclopedia Britannica*. https://www.britannica.com/science/parthenogenesis

32 Watts, P.C., Buley, K.R., Sanderson, S., Boardman, W., Ciofi, C. & Gibson, R. (2006). Parthenogenesis in Komodo Dragons. *Nature*, 444.

33 Shine, R. & Somaweera, R. (2019). Last Lizard Standing: The Enigmatic Persistence of the Komodo Dragon. *Global Ecology and Conservation*, 18.

34 Purwandana, D., Imansyah, M.J., Ariefiandy, A., Rudiharto, H., Ciofi, C. & Jessop, T.S. (2020). Insights into the Nesting Ecology and Annual Hatchling Production of the Komodo Dragon. *Copeia*, 108(4).

35 Birks, S.M. (1997). Paternity in the Australian brush-turkey, Alectura lathami, a megapode bird with uniparental male care. *Behavioral Ecology*, 8 (5). S.

36 San Diego Zoo Wildlife Alliance (et al.). *Australian Brush Turkey*. Accessed 25.11.22 https://animals.sandiegozoo.org/animals/australian-brush-turkey-0

37 Milius, S. (2000, 11.03). *Pregnant – and Still Macho*. Science News Online. http://ase.tufts.edu/biology/labs/lewis/news/articles/2000ScienceNews.pdf

38 Van Look, K., Dzyuba, B., Cliffe, A., Koldewey, H.J. & Holt, W.V. (2007). Dimorphic sperm and the unlikely route to fertilisation in the yellow seahorse. *J. Exp. Biol.*, 210(3).

39 Jones, A.G. (2004). Male pregnancy and the formation of seahorse species. *Biologist*, 51(4).

40 Holck, P. (2021). *Det gule legemet* i Store medisinske leksikon. https://sml.snl.no/det_gule_legemet (in Norwegian. Alternative source in English: Huffman, J. W. (2024, 24.04). *Pregnancy*. Encyclopedia Britannica. https://www.britannica.com/science/pregnancy).

41 Vestre, K. (2021). *The Making of You*. Profile/Wellcome Collection, tr. Matt Bagguley. (Originally published as *Det første mysteriet*. 2018. Aschehoug.)

42 Staff, Annetine, Professor of Obstetrics and Gynaecology at the University of Oslo. Personal communication, 05.11.22.

43 All the individuals in a species have the same set of genes, but have differing alleles. However, people commonly talk about genes when they really mean alleles, and there are a number of occasions in this book where I use the term gene instead of allele in order to simplify the language.

44 Haig, D. (1993). Genetic conflicts in human pregnancy. *Quarterly Review of Biology*, 68(4).

45 Schrödinger was a physicist who devised a thought experiment to show that the theory of quantum mechanics was

inconclusive – it's a long story, but in essence, there's a cat in a box, and there's a 50 per cent chance of it being dead owing to a radioactive particle. Theoretically, therefore, the cat is simultaneously alive and dead, but if you were to open the box, it would only be one of those things. It can feel that way to have an embryo inside you.

46 Hind, L.J. (2015, 9.11). *Ærfuglvokterne – naturens vaktmestere*. Forskning.no https://forskning.no/partner-naturvern-fugler/aerfuglevokterne–naturens-vaktmestere/459976 (In Norwegian. Alternative source in English: Brandslet, S. (2018, 01.03.) Eiderdown farming – a living cultural tradition. Norwegian SciTech News: https://norwegianscitechnews.com/2018/03/eider-farming-living-cultural-tradition/).

47 Dybdal, S.E. (2015, 6.07). *Dyne med norsk ærfugldun er verdas beste*. Nibio nyheter. https://www.nibio.no/nyheter/dyne-med-norsk-rfugldun-er-verdas-beste (In Norwegian. Alternative source in English: Noonan, M. L. (2018, 02.10) The world's lightest, warmest and most expensive down. BBC. https://www.bbc.com/travel/article/20181001-the-worlds-lightest-warmest-and-most-expensive-down).

48 Öst, M. & Bäck, A. (2003). Spatial structure and parental aggression in eider broods. *Animal Behaviour*, 66(6).

49 Friebe, A., Evans, A.L., Arnemo, J.M., Blanc, S., Brunberg, S., Fleissner, G., Swensson, J.E. & Zedrosser, A. (2014). Factors Affecting Date of Implantation, Parturition, and Den Entry Estimated from Activity and Body Temperature in Free-Ranging Brown Bears. *PLOS ONE*, 9(7).

50 UW Medicine (2020, 7.02). *How some mammals pause their pregnancies*. UW Medicine Newsroom. https://newsroom.uw.edu/news/how-some-mammals-pause-their-pregnancies

51 Teigland, S.C. (2017, 29.04). *For tynn for å bli gravid*. Klikk.no https://www.klikk.no/foreldre/gravid/for-tynn-til-a-bli-gravid-2360165. (In Norwegian. Alternative source in English:

The ESHRE Capri Workshop Group (2006, 05). Nutrition and reproduction in women. *Human Reproduction Update*, 12(3) https://academic.oup.com/humupd/article/12/3/193/554114)

52 San Diego Zoo Wildlife Alliance (2021, 10.08). *Platypus (Ornithorhynchus anatinus) Fact Sheet: Reproduction & Development.* https://ielc.libguides.com/sdzg/factsheets/platypus/reproduction

53 Bino, G., Kingsford, R.T., Archer, M., Connolly, J.H., Day, J., Dias, K., Goldney, D., Gongora, J., Grant, T., Griffiths, J., Hawke, T., Klamt, M., Lunney, D., Mijangos, L., Munks, S., Sherwin, W., Serena, M., Temple-Smith, P., Thomas, J., Williams, G. & Whittington, C. (2019). The platypus: evolutionary history, biology, and an uncertain future. *Journal of Mammalogy*, 100(2).

54 Castillo, M.A. & Kight, S.L. (2005). Response of terrestrial isopods, *Armadillidium vulgare* and *Porcellio laevis* (Isopoda: Oniscidea) to the ant *Tetramorium caespitum*: Morphology, behavior and reproductive success. *Invertebrate Reproduction and Development*, 47(3).

55 Vestre, K. (2021). *The Making of You*. Profile/Wellcome Collection, tr. Matt Bagguley. (Originally published as *Det første mysteriet*. 2018. Aschehoug.)

56 Wikipedia (n.d.). *Squamata*. Accessed 19.03.2024 at https://en.wikipedia.org/wiki/Squamata#Reproduction

57 Burrage, B. (1973). Comparative ecology and behavior of Chamaeleo pumilus pumilus (Gmelin) and C.namaquensis A. Smith (Sauria: Chamaeleonidae). *Annals of the South African Museum*, 61.

58 Tyndale-Biscoe, H. & Renfree, M. (1987). *Reproductive Biology of Marsupials*. Cambridge University Press.

59 Joo, M. (2004). *Macropus giganteus*. Animal Diversity Web. https://animaldiversity.org/accounts/Macropus_giganteus/

60 Tyndale-Biscoe, H. & Renfree, M. (1987). *Reproductive Biology of Marsupials*. Cambridge University Press.

61 Fenelon, J.C., Banerjee, A. & Murphy, B.D. (2014). Embryonic diapause: development on hold. *Int. J. Dev. Biol.* 58.
62 Flaxman, S.M. & Sherman, P.W. (2000). Morning sickness: a mechanism for protecting mother and embryo. *Quarterly Review of Biology*, 75(2).
63 Ibid.
64 Pepper, G.V. & Roberts, S.C. (2006). Rates of nausea and vomiting in pregnancy and dietary characteristics across populations. *Proc. Biol. Sci.* 273(1601).
65 Flaxman, S.M. & Sherman, P.W. (2008). Morning Sickness: Adaptive Cause or Nonadaptive Consequence of Embryo Viability? *The American Naturalist*, 172(1).
66 Gadsby, R., Ivanova, D., Trevelyan, E., Hutton, J.L. & Johnson, S. (2020). Nausea and vomiting in pregnancy is not just 'morning sickness': data from a prospective cohort study in the UK. *Br. J. Gen Pract.*, 70(697).
67 The time from egg-laying to hatching is approximately forty days. Bilde, Trine, Professor of Biology at Aarhus University. Personal communication 25.08.22.
68 Junghanns, A., Holm, C., Schou, M.F., Sørensen, A.B., Uhl, G. & Bilde, T. (2017). Extreme allomaternal care and unequal task participation by unmated females in a cooperatively breeding spider. *Animal Behaviour* 132.
69 Stenseth, N.C. (2021). *Slektskapsseleksjon* i Store norske leksikon. https://snl.no/slektskapsseleksjon. (In Norwegian. Alternative source in English: Britannica, T. Editors of Encyclopaedia (2018, 14.04). *Kin selection. Encyclopedia Britannica.* https://www.britannica.com/topic/kin-selection).
70 Høiland, K. (2018). *Snylteklubbe* i Store norske leksikon. https://snl.no/snylteklubbe. In Norwegian. Alternative source in English: Wikipedia. (n.d.). Cordyceps. Retrieved 02.05.2024, from https://en.wikipedia.org/wiki/Cordyceps).

71 Trevathan, W.R. & Rosenberg, K.R. (2020). Evolutionary Medicine and Women's Reproductive Health in Schulkin, J. & Power, M.L., *Integrating Evolutionary Biology into Medical Education—for maternal and child healthcare students, clinicians, and scientists*. Oxford University Press.

72 Natural selection is the more important evolutionary mechanism, but evolution can also occur via gene flow and genetic drift. Those interested in further reading are referred to the page "Evolution" in Encyclopaedia Britannica: https://www.britannica.com/science/evolution-scientific-theory.

73 Laidlaw, S. (2020). *Giant Pacific Octopus* in Biology Dictionary. https://biologydictionary.net/giant-pacific-octopus/

74 Wood, J.B. (n.d.). *Enteroctopus dofleini, The Giant Pacific Octopus* on The Cephalopod Page. Retrieved 30.11.22, from http://www.thecephalopodpage.org/Edofleini.php

75 Blaas, H-G.K. (2017). Embryoets og fosterets utvikling, in Brunstad, A. & Tegnander, E., *Jordmorboka – ansvar, funksjon og arbeidsområde*. Cappelen Damm Akademisk. (In Norwegian. Alternative source in English: Gai, Z., Zhu, M., Ahlberg, P.E., & Donoghue, P.C.J. (2022) The Evolution of the Spiracular Region From Jawless Fishes to Tetrapods, in Frontiers in Ecology and Evolution 2022 (10) https://www.frontiersin.org/articles/10.3389/fevo.2022.887172).

76 Grunstra, N.D.S., Zachos, F.E., Herdina, A.N., Fischer, B., Pavličev, M. & Mitteroecker, P. (2019). Humans as inverted bats: A comparative approach to the obstetric conundrum. *American Journal of Human Biology* 31:e23227.

77 Wikipedia (n.d.). *Emperor penguin – Courtship and breeding*. Retrieved 17.04.2022 from https://en.wikipedia.org/wiki/Emperor_penguin#Courtship_and_breeding

78 Folch, A. (1992). Family Apterygidae (Kiwis) in del Hoyo, J., Elliot, A. & Sargatal, J. (ed.), *Handbook of the Birds of the World*, (Vol. 1). Lynx Edicions.

79 Save the Kiwi. *Producing an egg*. Retrieved 11.05.22 from https://www.savethekiwi.nz/about-kiwi/kiwi-facts/kiwi-life-cycle/

80 Abourachid, A., Castro, I. & Provini, P. (2019). How to walk carrying a huge egg? Trade-offs between locomotion and reproduction explain the special pelvis and leg anatomy in kiwi (Aves; Apteryx spp.) *Journal of Anatomy*, 235.

81 Folch, A. (1992). Family Apterygidae (Kiwis) in del Hoyo, J., Elliot, A. & Sargatal, J. (eds.), *Handbook of the Birds of the World*, (Vol. 1). Lynx Edicions.

82 Dean, S. (2015). *Why is the Kiwi's Egg So Big?* Audubon. https://www.audubon.org/news/why-kiwis-egg-so-big

83 The gestation period lasts for approximately seventy days or more, varying significantly with environmental factors, temperature in particular. cf. Greven, H., Flossdorf, D., Köthe, J., List, F. & Zwanzig, N. (2014). Running Speed and Food Intake of the Matrotrophic Viviparous Cockroach Diploptera punctata (Blattodea: Blaberidae) during Gestation. *Entomologie Heute*, 26.

84 Or perhaps they don't: the Southern Darwin's frog is a severely endangered species in sharp decline.

85 Wikipedia (n.d.). *Darwin's frog*. Retrieved 05.09.22 from https://en.wikipedia.org/wiki/Darwin's_frog

86 Goicoechea, O., Garrido, O. & Jorquera, B. (1986). Evidence for a Trophic Paternal-Larval Relationship in the Frog Rhinoderma darwinii. *Journal of Herpetology*, 20(2).

87 Wikipedia. (n.d.) *Wandering albatross*. Retrieved 06.09.22 from https://en.wikipedia.org/wiki/Wandering_albatross

88 Nanda, S. (2000). *Gender Diversity. Crosscultural Variations*. Waveland Press.

89 Mating types is a strategy to avoid mating with oneself, and is equivalent to sex, but not directly comparable. All the same, it shows us that it's possible to imagine other ways of organising mating than producing large and small sex cells.

90 We distinguish between eukaryote organisms (those whose cells have a cell nucleus) and prokaryote (those whose cells do not). Prokaryote organisms cover the kingdoms bacteria and archaea, while the three kingdoms plants, fungi and animals are eukaryotes.

91 Otto, S. (2008). Sexual Reproduction and the Evolution of Sex. *Nature Education* 1(1).

92 Johnson, J.D., White, N.L., Kangabire, A. & Abrams, D.M. (2021). A dynamical model for the origin of anisogamy. *Journal of Theoretical Biology*, 521.

93 Fuentes, A. (2022). *Race, Monogamy, and Other Lies They Told You. Busting Myths About Human Nature*, (2nd edn.). University of California Press.

94 It's not always a good strategy for either the female or the male to mate with more than one partner – if it reduces investment in the offspring or their survival rates. This will vary from species to species.

95 Bagemihl, B. (2000). *Biological Exuberance: Animal Homosexuality and Natural Diversity*. Stonewall Inn Editions.

96 See Cooke, L. (2022). *Bitch: A Revolutionary Guide to Sex, Evolution and the Female Animal*. Transworld Publishers, for a description of how Gowaty and other women in science including Jeanne Altmann, Mary Jane West-Eberhard and Sarah Blaffer Hrdy have quashed existing ideas about sex and evolution.

97 Fine, C. (2017). *Testosterone Rex*. Icon Books.

98 Kokko, H. & Jennions, M.D. (2008). Parental investment, sexual selection and sex ratios. *Journal of Evolutionary Biology*, 21.

99 Liker, A., Freckleton, R.P., Remes, V. & Székely, T. (2015). Sex differences in parental care: Gametic investment, sexual selection, and social environment. *Evolution* 69(11).

100 Kokko, H. & Jennions, M.D. (2008). Parental investment, sexual selection and sex ratios. *Journal of Evolutionary Biology*, 21.

101 Auld, S.K.J.R, Tinkler, S.K. & Tinsley, M.C. (2016). Sex as a strategy against rapidly evolving parasites. *Proc. R. Soc B*, 283(1845).

102 John Maynard Smith pointed out "the two-fold cost of sex" in the 1970s. His model predicts that the per-capita birth rate of an asexual population would be twice that of a sexual population. Why sex exists is still a biological mystery.

103 Zhang, Y.-N., Zhu, X.-Y., Wang, W.-P., Wang, Y., Wang, L., Xu, X.-X., Zhang, K. & Deng, D.G. (2016). Reproductive switching analysis of Daphnia similoides between sexual female and parthenogenetic female by transcriptome comparison. *Sci Rep*, 6.

104 Auld, S.K.J.R, Tinkler, S.K. & Tinsley, M.C. (2016). Sex as a strategy against rapidly evolving parasites. *Proc. R. Soc. B*, 283(1845).

105 Barton, N.H. & Charlesworth, B. (1998). Why Sex and Recombinations? *Science*, 281(5385).

106 Casas, L., Saborido-Rey, F., Ryu, T., Michell, C., Ravasi, T. & Irigoien, X. (2016). Sex Change in Clownfish: Molecular Insights from Transcriptome Analysis. *Scientific Reports*, 6.

107 Schärer, L. & Ramm, S.A. (2016). Hermaphrodites in Kliman, R. (ed.), *Encyclopedia of Evolutionary Biology*, Vol. 2. Elsevier.

108 Kokko, H. & Jennions, M.D. (2008). Parental investment, sexual selection and sex ratios. *Journal of Evolutionary Biology*, 21.

109 Stenseth, N.C. & Voje, K. (2022). *Seksuell seleksjon* i Store norske leksikon. https://snl.no/seksuell_seleksjon (In Norwegian. Alternative source in English: Ayala, F. Jose (20.03.24). *Sexual selection. Encyclopedia Britannica*. https://www.britannica.com/science/sexual-selection).

110 Kokko, H. & Jennions, M.D. (2008). Parental investment, sexual selection and sex ratios. *Journal of Evolutionary Biology*, 21.

111 Cooke, L. (2022). *Bitch: A Revolutionary Guide to Sex, Evolution and the Female Animal*. Transworld Publishers.

112 Blackless, M., Charuvastra, A., Derryck, A., Fausto-Sterling, A., Lauzanne, K. & Lee, E. (2000). How sexually dimorphic are we? Review and synthesis. *American Journal of Human Biology*, 12(2).

113 Sørlie, A. (n.d.) *Hva er kjønnsinkongruens?* Retrieved 20.09.22 from https://kjonnsinkongruens.no/kjonnsinkongruens/ (In Norwegian. Alternative source in English: World Health Organization (n.d.). Gender incongruence and transgender health in the ICD. Retrieved 09.05.24 from https://www.who.int/standards/classifications/frequently-asked-questions/gender-incongruence-and-transgender-health-in-the-icd).

114 Hogenboom, M. (2021). *The gender biases that shape our brains*. BBC Future. https://www.bbc.com/future/article/20210524-the-gender-biases-that-shape-our-brains

115 Yong, E. (2013). *The Alligator Has a Permanently Erect, Bungee Penis*. National Geographic. https://www.nationalgeographic.com/science/article/the-alligatorhas-a-permanently-erect-bungee-penis

116 Brennan, P.L.R. & Orbach, D.N. (2020). Copulatory behavior and its relationship to genital morphology. *Advances in the Study of Behavior*, 52.

117 See sources referenced in Tavalieri, Y.E., Galoppo, G.H., Canesini, G., Truter, J.C., Ramos, J.G., Luque, EH. & Muñoz-de-Toro, M. (2019). The external genitalia in juvenile Caiman latirostris differ in hormone sex determinate-female from temperature sex determinate-female. *General and Comparative Endocrinology*, 273.

118 Kofron, C.P. (1989). Nesting ecology of the Nile crocodile (Crocodylus niloticus). *African Journal of Ecology*, 27.

119 Combrink, X., Warner, J.K. & Downs, C.T. (2017). Nest-site selection, nesting behaviour and spatial ecology of female Nile crocodiles (Crocodylus niloticus) in South Africa. *Behavioural Processes*, 135.

120 Depending on temperature, the eggs take 84–98 days to hatch, equivalent to twelve to fourteen weeks. See Combrink, et al. (2017), above.

121 Combrink, X., Warner, J.K. & Downs, C.T. (2016). Nest predation and maternal care in the Nile crocodile (Crocodylus niloticus) at Lake St Lucia, South Africa. *Behavioural Processes*, 133.

122 Lang, J. W. & Andrews, H.V. (1994). Temperature-dependent sex determination in crocodilians. *Journal of Experimental Zoology*, 270(1).

123 Shine, R. (1999). Why is sex determined by nest temperature in so many reptiles? Review. *Trends in Ecology and Evolution*, 14(5).

124 Spencer, R.-J. & Janzen, F.J. (2014). A novel hypothesis for the adaptive maintenance of environmental sex determination in a turtle. *Proc. Biol. Sci*, 281(1789).

125 Jarvis, J.U.M. & Bennett, N.C. (1991). The Ecology and Behavior of the Family Bathyergidae, in Sherman, P.W., Jarvis, J.U.M., & Alexander, R.D, *The Biology of the Naked Mole-Rat*. Princeton University Press.

126 Bennet, N.C. & Jarvis, J.U.M. (2004). Cryptomys damarensis. *Mammalian Species*, 756.

127 Haugaasen, J.M.T., Haugaasen, T., Peres, C.A., Gribel, R. & Wegge, P. (2010). Seed dispersal of the Brazil nut tree (Bertholletia excelsa) by scatter-hoarding rodents in a central Amazonian forest. *Journal of Tropical Ecology*, 26.

128 Juni, E. (2011). *Myoprocta pratti*. Animal Diversity Web. https://animaldiversity.org/accounts/Myoprocta_pratti/

129 Holck, P. (2022). *Skjedekrans* i Store norske leksikon. https://sml.snl.no/skjedekrans (in Norwegian. Alternative source in English: Britannica, T. Editors of Encyclopaedia (26.02.24). *Hymen. Encyclopedia Britannica*. https://www.britannica.com/science/hymen-anatomy).

130 Freeman, A.R. (2021). Female–Female Reproductive

Suppression: Impacts on Signals and Behavior. *Integrative and Comparative Biology*, 61(5).

131 Swift, J. (2021). *Looking for love, finding TNT.* https://as.cornell.edu/news/looking-love-finding-tnt

132 Juni, E. (2011). *Myoprocta pratti.* Animal Diversity Web. https://animaldiversity.org/accounts/Myoprocta_pratti/

133 Admittedly, the common shrew is not pregnant for fifteen weeks, she's pregnant for a little more than three weeks. She reproduces continuously throughout the summer, eating non-stop all the while.

134 Frafjord, K. (2022). *Spissmus* i Store norske leksikon. https://snl.no/spissmus (in Norwegian. Alternative source in English: Musser, G. (12.04.24). *Shrew.* Encyclopedia Britannica. https://www.britannica.com/animal/shrew).

135 Rossell, F. & Pedersen, K.V. (1999). *Bever.* Landbruksforlaget. (In Norwegian. Alternative source in English: Musser, G. (27.04.24). *Beaver.* Encyclopedia Britannica. https://www.britannica.com/animal/beaver).

136 Kruuk, H. (1972). *The spotted hyena. A study of Predation and Social Behavior.* The University of Chicago Press.

137 Bondar, C. (2016, 26.08). *For Some Species, the Girls Come with Boy Bits.* PBS Nature. https://www.pbs.org/wnet/nature/blog/girls-boy-bits-pseudo-penis-hyena-elephant/

138 Wilke, C. (2020, 25.08). *Female hyenas kill off cubs in their own clans.* Science News. https://www.sciencenews.org/article/female-hyena-moms-kill-cubs-own-clans

139 Frank, L.G. & Glickman, S.E. (1994). Giving birth through a penile clitoris: parturition and dystocia in the spotted hyena (Crocuta crocuta) *J. Zool. Lond.*, 234.

140 Glickman, S.E., Cunha, G.R., Drea, C.M., Conley, A.J. & Place, N.J. (2006). Mammalian sexual differentiation: lessons from the spotted hyena. *Trends in Endocrinology and Metabolism*, 17(9).

141 Frank, L.G. & Glickman, S.E. (1994). Giving birth through a penile clitoris: parturition and dystocia in the spotted hyena (Crocuta crocuta) *J. Zool. Lond.*, 234.

142 Wilke, C. (2020, 25.08). *Female hyenas kill off cubs in their own clans.* Science News. https://www.sciencenews.org/article/female-hyena-moms-kill-cubs-own-clans

143 Cunha, G.R., Risbridger, G., Wang, H., Place, N.J., Grumbach, M., Cunha, T.J., Weldele, M., Conley, A.J., Barcellos, D., Agarwal, S., Bhargava, A., Drea, C., Hammond, G.L., Siiteri, P., Coscia, E.M., McPhaul, M.J., Baskin, L.S. & Glickman, S.E. (2014). Development of the external genitalia: Perspectives from the spotted hyena (Crocuta crocuta). *Differentiation*, 87.

144 See Gross, R.E. (2022). *Vagina Obscura: an anatomical voyage.* W.W. Norton & Company, for the human clitoris, Cooke, L. (2022). *Bitch. A revolutionary guide to sex, evolution & the female animal.* Transworld Publishers, for clitoris research in general, and Folwell, M., Sanders, K. & Crowe-Riddell, J. (2022). The Squamate Clitoris: A Review and Directions for Future Research. *Integrative and Comparative Biology*, 62(3) for squamata clitorises.

145 Tronstad, T.T. (2021), Stillhetens klitoris. *Samtiden*, 3. (In Norwegian. Alternative source in English: Gross, Rachel E. (2022) Vagina Obscura: an anatomical voyage. W.W. Norton & Company)

146 Gross, R. E. (2022). *Vagina Obscura: an anatomical voyage.* W. W. Norton & Company.

147 Putka, S. (2022, 04.11). *8000 Nerve Endings? Actually, the Clitoris Has More.* MedPage Today. https://www.medpagetoday.com/meetingcoverage/smsna/101464 (The study was led by Blair Peters, Assistant Professor of Surgery at Oregon Health & Science University, who specialises in gender-affirming surgery on transmasculine people).

148 Jowitt, M. (2018). The Clitoris in Labor. *Midwifery Today*, 127.

149 Ortega, J. & Alarcón-D., I. (2008). Anoura geoffroyi (Chiroptera: Phyllostomidae). *Mammalian Species*, 818.

150 Grunstra, N.D.S., Zachos, F.E., Herdina, A.N., Fischer, B., Pavličev, M. & Mitteroecker, P. (2019). Humans as inverted bats: A comparative approach to the obstetric conundrum. *American Journal of Human Biology* 31:e23227.

151 Grunstra, N.D.S., Zachos, F.E., Herdina, A.N., Fischer, B., Pavličev, M. & Mitteroecker, P. (2019). Humans as inverted bats: A comparative approach to the obstetric conundrum. *American Journal of Human Biology* 31:e23227.

152 Ibid.

153 Quinlan, K.C. (2021). *San Martin Titi*. New England Primate Conservancy. https://neprimateconservancy.org/san-martin-titi/

154 This San Martin titi birth account is based on an observation in the wild by Dr. Anneke M. DeLuycker and her field assistant Rosse Mary Vásquez Ríos, referenced in DeLuycker, A.M. (2014). Observations of a daytime birthing event in wild titi monkeys (*Callicebus oenanthe*): implications of the male parental role. *Primates*, 55. A single delivery does not necessarily represent the standard delivery in a species, and the length of delivery, the point at which the male provides care to the newborn, and other events may vary significantly.

155 Ibid.

156 Hrdy, S.B. (2011). *Mothers and Others. The evolutionary origins of mutual understanding.* The Belknap Press.

157 Frogs and toads are the two families that make up the order tailless amphibians (Anura).

158 Recorded incubation times vary from eleven to twenty-two weeks in the Surinam toad, probably depending on temperature. See Zippel, K.C. (2006). Further observations of oviposition in the Surinam toad (Pipa pipa), with comments on biology, misconceptions, and husbandry. *Herpetological Review*, 37.

159 Fernandes T.L., Antoniazzi, M.M., Sasso-Cerri, E., Egami, M.I.,

Lima, C., Rodrigues, M.T. & Jared, C. (2011). Carrying Progeny on the Back: Reproduction in the Brazilian Aquatic Frog Pipa carvalhoi. *South American Journal of Herpetology*, 6(3).

160 Zippel, K.C. (2006). Further observations of oviposition in the Surinam toad (Pipa pipa), with comments on biology, misconceptions, and husbandry. *Herpetological Review*, 37.

161 Ibid.

162 Burrage, B. (1973). Comparative ecology and behavior of Chamaeleo pumilus pumilus (Gmelin) and C. namaquensis A. Smith (Sauria: Chamaeleonidae). *Annals of the South African Museum*. 61.

163 Buer, H. (2011). *Villsauboka*. Selja forlag. (In Norwegian. The title translates as 'The Wild Sheep Book').

164 Wikipedia. (n.d.). *Oestrus ovis*. Retrieved 01.12.22 from https://en.wikipedia.org/wiki/Oestrus_ovis

165 Bainbridge, D. (2001). *Making babies. The science of pregnancy.* Harvard University Press.

166 Nesheim, B.-I. (2022). *Morkaken* i Store norske leksikon. https://sml.snl.no/morkaken (in Norwegian. Alternative source in English: Huffman, J.W. (09.05.24). *pregnancy. Encyclopedia Britannica.* https://www.britannica.com/science/pregnancy).

167 Wang, Z.Y. & Ragsdale, C.W. (2018). Multiple optic gland signaling pathways implicated in octopus maternal behaviors and death. *J. Exp. Biol.*, 221(19).

168 Young, T.P. (2010). Semelparity and Iteroparity. *Nature Education Knowledge*, 3(10).

169 Yong, E. (2013). *Why A Little Mammal Has So Much Sex That It Disintegrates.* National Geographic. https://www.nationalgeographic.com/science/article/why-a-little-mammal-has-so-much-sex-that-it-disintegrates

170 See Kindsvater et al. (2016), cited in Wang, Z.Y. & Ragsdale, C.W. (2018). Multiple optic gland signaling pathways implicated in octopus maternal behaviors and death. *J. Exp. Biol.*, 221(19).

171 See Wodinsky, J. (1977), cited in Wang, Z.Y. & Ragsdale, C.W. (2018). Multiple optic gland signaling pathways implicated in octopus maternal behaviors and death. *J. Exp. Biol.*, 221(19).

172 Wang, Z. (2018). *Molecular Neuroendocrinology of Maternal Behaviors and Death in the California Two-Spot Octopus, Octopus bimaculoides* [Doctoral thesis]. The University of Chicago.

173 Benirsche, K. (2007). *Thomson's Gazelle* in Comparative Placentation. http://placentation.ucsd.edu/thom.htm

174 Costelloe, B.R. & Rubenstein, D.I. (2015). Coping with transition: offspring risk and maternal behavioural changes at the end of the hiding phase. *Animal Behaviour*, 109.

175 Oh, how naive I was!

176 Purser, A., Hehemann, L., Boehringer, L., Tippenhauer, S., Wege, M., Bornemann, H., Pineda-Metz, S.E.A., Flintrop, C.M., Koch, F., Hellmer, H.H., Burkhardt-Holm, P., Janout, M., Werner, E., Glemser, B., Balaguer, J., Rogge, A., Holtappels, M. & Wenzhoefer, F. (2022). A vast icefish breeding colony discovered in the Antarctic. *Current Biology*, 32.

177 Riginella, E., Pineda-Metz, S.E.A., Gerdes, D., Koschnick, N., Bömer, A., Biebow, H., Papetti, C., Mazzoldi, C. & La Mesa, M. (2021). Parental care and demography of a spawning population of the channichthyid Neopagetopsis ionah, Nybelin 1947 from the Weddell Sea. *Polar Biology*, 44.

178 National Geographic (2022). Ocean. Resource library. https://education.nationalgeographic.org/resource/ocean

179 Kock, K., & Kellermann, A. (1991). Reproduction in Antarctic notothenioid fish. *Antarctic Science*, 3(2).

180 Horner, J.R. & Currie, P.J. (1994) Embryonic and neonatal morphology and ontogeny of a new species of *Hypacrosaurus* (Ornithischia, Lambeosauridae) from Montana and Alberta in Carpenter, K., Hirsch, K.F. & Horner, J.R. (eds.), *Dinosaur Eggs and Babies*. Cambridge University Press.

181 Erickson, G.M., Zelenitsky, D.K., Kay, D.I. & Norell, M.A.

(2017) Dinosaur incubation periods directly determined from growth-line counts in embryonic teeth show reptilian-grade development. *PNAS*, 114(3).

182 Tanaka, K., Zelenitsky, D., Therrien, F. & Kobayashi, Y. (2018). Nest substrate reflects incubation style in extant archosaurs with implications for dinosaur nesting habits. *Scientific Reports*, 8:3170.

183 Erickson, G.M., Zelenitsky, D.K., Kay, D.I. & Norell, M.A. (2017). Dinosaur incubation periods directly determined from growth-line counts in embryonic teeth show reptilian-grade development. *PNAS*, 114(3).

184 Horner, J.R. & Currie, P.J. (1994) Embryonic and neonatal morphology and ontogeny of a new species of *Hypacrosaurus* (Ornithischia, Lambeosauridae) from Montana and Alberta in Carpenter, K., Hirsch, K.F. & Horner, J.R. (eds.), *Dinosaur Eggs and Babies*. Cambridge University Press.

185 Cooper, L.N., Lee, A.H., Taper, M.L. & Horner, J.R. (2008). Relative growth rates of predator and prey dinosaurs reflect effects of predation. *Proc. Biol. Sci.*, 275(1651).

186 Dawson, J. (2014). *Egg Mountain, the Two Medicine, and the Caring Mother Dinosaur*. National Park Service. https://www.nps.gov/articles/mesozoic-egg-mountaindawson-2014.htm

187 Horner, J.R. & Makela, R. (1979). Nest of juveniles provides evidence of family structure among dinosaurs. *Nature*, 282(5736).

188 Black, R. (2020, 24.07). *How Dinosaurs Raised Their Young*. Smithsonian Magazine. https://www.smithsonianmag.com/science-nature/dinosaurs-parents-new-egg-discovery-180975361/

189 Symeou, A. *8 Facts You (probably) Didn't Know About Sloths' Anatomy*. The Sloth Conservation Foundation. Accessed 28/11/22 https://slothconservation.org/8-facts-about-sloths-skeleton-anatomy/

190 Gilmore, D.P., Da-Costa, C.P. & Duarte, D.P.F. (2000). An update on the physiology of two- and three-toed sloths. *Brazilian Journal of Medical and Biological Research*, 33.

191 Taube, E., Keravec, J., Vié, J.-C. & Duplantier, J.-M. (2001). Reproductive biology and postnatal development in sloths, Bradypus and Choloepus: review with original data from the field (French Guiana) and from captivity. *Mammal Review*, 31.

192 Some species of two-toed sloth are reported to give birth both suspended and on the ground, but the brown-throated sloth (which is a species of three-toed sloth) only gives birth in the canopy.

193 Sverdrup-Thygeson, A. (2020). *Dovendyret og sommerfuglen*. Ena forlag. (In Norwegian. Children's picture book. The title translates as: "The Sloth and the Butterfly".)

194 Pauli, J.N., Mendoza, J.E., Steffan, S.A., Carey, C.C., Weimer, P.J. & Peery, M.Z. (2014). A syndrome of mutualism reinforces the lifestyle of a sloth. *Proc. R. Soc. B.*, 218(1778).

195 The wolffish in the Saltstraum maelstrom are recorded as guarding the egg ball for around seven weeks before the eggs hatch (Karlsen, Vebjørn, diver. Personal correspondence 27.05.22), while other sources state an incubation period of six and ten months in the same species in other locations in the world.

196 Hrdy, S.B. (2011). *Mothers and Others. The evolutionary origins of mutual understanding.* The Belknap Press.

197 Hrdy, S.B. (1986). Empathy, Polyandry, and the Myth of the Coy Female in Bleier, R. (red.) *Feminist Approaches to Science.* Pergamon Press.

198 Hrdy, S.B. (2011). *Mothers and Others. The evolutionary origins of mutual understanding.* The Belknap Press.

199 Caro, S.M., Griffin, A.S., Hinde, C.A. & West, S.A. (2016). Unpredictable environments lead to the evolution of parental neglect in birds. *Nature Communications*, 7:10985.

200 U.S. Fish and Wildlife Service Pacific Islands. 08.12.22. *Dropping Some Wisdom On You!* Facebook. https://www.facebook.com/PacificIslandsFWS

201 Lahdenperä, M., Mar, K.U. & Lummaa, V. (2016). Nearby grandmother enhances calf survival and reproduction in Asian elephants. *Scientific Reports*, 6:27213.

202 Stansfield, F.J., Nöthling, J.O. & Allen, W.R. (2013). The progression of small-follicle reserves in the ovaries of wild African elephants (Loxodonta africana) from puberty to reproductive senescence. *Reprod. Fertil Dev.*, 25(8).

203 Uematsu, K., Kutsukake, M., Fukatsu, T., Shimada, M. & Shibao, H. (2010). Altruistic Colony Defense by Menopausal Female Insects. *Current Biology*, 20.

204 FN-sambandet (2020). *Barnedødelighet.* https://www.fn.no/Statistikk/barnedoedelighet (in Norwegian. Alternative source in English: UN IGME (2024) Levels and trends in child mortality. https://data.unicef.org/resources/levels-and-trends-in-child-mortality-2024/).

205 Blell, M. (2018). Grandmother Hypothesis, Grandmother Effect, and Residence Patterns, in Callan, H. (ed.), *The International Encyclopedia of Anthropology.* John Wiley & Sons.

206 Hawkes, K., O'Connell, J.F., Blurton Jones, N.G., Alvarez, H. & Charnov, E.L. (1998). Grandmothering, menopause, and the evolution of human life histories. PNAS, 95(3). S.1336–1339. I also used Saini, A. (2017). *Inferior: How Science Got Women Wrong – and the New Research That's Rewriting the Story.* Beacon Press, and Cooke, L. (2022). *Bitch: A revolutionary guide to sex, evolution & the female animal.* Transworld Publishers, as background material for this chapter.

207 See Saini, A. (2017). *Inferior: How Science Got Women Wrong – and the New Research That's Rewriting the Story.* Beacon Press, for a summary.

208 Croft, D.P., Brent, L.J.N., Franks, D.W. & Cant, M.A. (2015). The evolution of prolonged life after reproduction. *Trends in Ecology and Evolution*, 30(7).

209 There are two distinct ecotypes of orca in this region: those who live in large pods along the coast and eat fish ("residents"), and those who live in small groups of two to six animals further out to sea, eating seals and other mammals ("transients"). There are also ecotypes who live even further out in the ocean (and are therefore harder to study). The two ecotypes communicate and hunt differently, and do not mix. When I refer to orca here, the information is based on "residents" along the west coast of the USA and Canada.

210 Natrass, S., Croft, D.P., Ellis, S., Cant, M.A., Weiss. M.N., Wright, B.M., Stredulinsky, E., Doniol-Valcroze, T., Ford, J.B.K., Balcomb, K.C. & Franks, D.W. (2019). Postreproductive killer whale grandmothers improve the survival of their grandoffspring. *PNAS*, 116(52).

211 Croft, D.P., Johnstone, R.A., Ellis, S., Nattrass, S., Franks, D.W., Brent, L.J.N., Mazzi, S., Balcomb, K.C., Ford, J.K.B. & Cant, M.A. (2017). Reproductive Conflict and the Evolution of Menopause in Killer Whales. *Current Biology*, 27.

212 Johnstone, R.A. & Cant, M.A. (2010). The evolution of menopause in cetaceans and humans: the role of demography. *Proc. R. Soc. B*, 277.

213 Lin, C.-H., Takahashi, S., Mulla, A.J. & Nozawa, Y. (2021). Moonrise timing is key for synchronized spawning in coral Dipsastraea speciosa. *PNAS*, 118(34).

214 Anderson, M.V, & Rutherford, M.D. (2013). Evidence of a nesting psychology during human pregnancy. *Evolution and Human Behavior*, 34.

215 Eurostat (2019). *Participation time per day in household and family care, by gender.* https://ec.europa.eu/eurostat/statistics-explained/index.

php?title=File:Participation_time_per_day_in_household_and_family_care,_by_gender,_(hh_mm;_2008_to_2015).png

216 Shavisi, A. (2020). Nesting behaviours during pregnancy: Biological instinct, or another way of gendering housework? *Women's Studies International Forum*, 78:102329.

217 See Fuentes, A. (2022). *Race, Monogamy and Other Lies They Told You*. University of California Press, for a good overview of the most common myths about human nature.

218 Tveter, N. (2016, 14/12). *Den magiske reinsdyrnesen*. Gemini.no. https://gemini.no/2016/12/den-magiske-reinsdyrnesen/ (in Norwegian. Alternative source in English: Tveter, N. (15.12.16) The magical reindeer nose. Retrieved 13.05.24 from https://www.sintef.no/en/latest-news/2016/the-magical-reindeer-nose/).

219 Holand, Ø. & Punsvik, T. (2016) Villreinen – en suksessfull art, in Punsvik, T. and Frøstrup, J.C. (eds.), *Fjellviddas nomade – Villreinen*. Friluftsforlaget. (In Norwegian. Alternative source in English: Britannica, T. Editors of Encyclopaedia (12.04.24). reindeer. Encyclopedia Britannica. https://www.britannica.com/animal/reindeer).

220 Strand, O. & Hansen, F.K. (2015). *Midt i flokken*. Kom forlag. (In Norwegian. Alternative source in English: Britannica, T. Editors of Encyclopaedia (12.04.24). reindeer. Encyclopedia Britannica. https://www.britannica.com/animal/reindeer).

221 Åsbakk, K. & Nilssen, A.C. (2014). Reinens hudbrems og svelgbrems: biologi, betydning og om bekjempelsestiltak. *Norsk veterinærtidsskrift*, 2. (in Norwegian. Alternative source in English: Britannica, T. Editors of Encyclopaedia (24.01.24). bot fly. Encyclopedia Britannica. https://www.britannica.com/animal/bot-fly)

222 Fischer, B., Grunstra, N.D.S., Zaffarini, E. & Mitteroecker, P. (2021). Sex differences in the pelvis did not evolve de novo in modern humans. *Nature Ecology & Evolution*, 5.

223 Grunstra, N.D.S., Zachos, F.E., Herdina, A.N., Fischer, B.,

Pavličev, M. & Mitteroecker, P. (2019). Humans as inverted bats: A comparative approach to the obstetric conundrum. *American Journal of Human Biology* 31:e23227.

224 Mitteroecker, P. & Fischer, B. (2023). Evolution of the human birth canal. *American Journal of Obstetrics & Gynecology.*

225 Grunstra, N.D.S., Zachos, F.E., Herdina, A.N., Fischer, B., Pavličev, M. & Mitteroecker, P. (2019). Humans as inverted bats: A comparative approach to the obstetric conundrum. *American Journal of Human Biology* 31:e23227.

226 That is, the ancestors of the amniotes, species who have membranes around their eggs.

227 Fischer, B., Grunstra, N.D.S., Zaffarini, E. & Mitteroecker, P. (2021). Sex differences in the pelvis did not evolve de novo in modern humans. *Nature Ecology & Evolution*, 5.

228 Voje, K.L. (2022). *Livets evolusjonshistorie* i Store norske leksikon. https://snl.no/livets_evolusjonshistorie (in Norwegian. Alternative source in English: Ayala, F. Jose (23.04.24). *evolution. Encyclopedia Britannica*. https://www.britannica.com/science/evolution-scientific-theory).

229 Frontiers in Ecology and Evolution (no date). *Origin and Early Evolution of Amniotes*. Research Topic. Accessed 15/09/22 https://www.frontiersin.org/research-topics/14947/origin-and-early-evolution-of-amniotes

230 Eltringham, S.K. (1999). *The Hippos*. Academic Press.

231 Mason, K. (2013). *Hippopotamus amphibius*. Animal Diversity Web. https://animaldiversity.org/accounts/Hippopotamus_amphibius/

232 Eltringham, S.K. (1999). *The Hippos*. Academic Press.

233 Lewinson, R. (1998). Infanticide in the hippopotamus: evidence for polygynous ungulates. *Ethology Ecology & Evolution*, 10.

234 Different sources state different hatching times, between seven and nine months. As with other reptiles, the incubation period is dependent on temperature and other environmental factors.

235 Komodo Survival Program (n.d.). *Life History*. Retrieved 05.07.22 from https://komododragon.org/life-history/
236 Ivančić, M., Gomez, F.M., Musser, W.B., Barratclough, A., Meegan, J.M., Waitt, S.M., Llerenas, A.C., Jensen, E.D. & Smith, C.R. (2020). Ultrasonographic findings associated with normal pregnancy and fetal well-being in the bottlenose dolphin (Tursiops truncatus). *Vet Radiol Ultrasound.*, 61(2). pp. 215–26.
237 Chose, T. (2013, 22/07). *Hey Flipper! Dolphins Use Names to Reunite.* LiveScience. https://www.livescience.com/38343-dolphin-whistles-act-like-names.html
238 Ames, A.E., Macgregor, R.P., Wielandt, S.J., Cameron, D.M., Kuczaj II, S.A. & Hill, H.M. (2019). Pre- and Post-Partum Whistle Production of a Bottlenose Dolphin (Tursiops truncatus) Social Group. *International Journal of Comparative Psychology*, 32.
239 This has been demonstrated in some studies, though there are other studies that have not demonstrated it. Therefore, there is no conclusive evidence that all dolphins sing their names to their young in utero.
240 Carvalho, M.E., Justo, J.M.R. de M., Gratier, M., da Silva, H.M.F.R. (2018). The Impact of Maternal Voice on the Fetus: A Systematic Review. *Current Women's Health Reviews*, 14(3).
241 Bellem, A.C., Monford, S.L. & Goodrowe, K.L. (1995). Monitoring reproductive development, menstrual cyclicity, and pregnancy in the lowland gorilla (Gorilla gorilla) by enzyme immunoassay. *Journal of Zoo and Wildlife Medicine*, 26(1).
242 Stewart, K.J. (1984). Parturition in Wild Gorillas: Behaviour of Mothers, Neonates, and Others. *Folia primatol.*, 42.
243 A video of the gorilla Calaya's birth at the Smithsonian's National Zoo on 15.04.2018 (from Youtube: https://www.youtube.com/watch?v=i497TV5Q6TY). NB: a single birth account is not indicative of how all gorillas give birth, and particularly not births in captivity. The birthing position, how the newborn is handled after the birth, and so on, will vary.

244 Rosenberg, K.R. & Trevathan, W.R. (2001). The Evolution of Human Birth. *Sci. Am.*, 285 (5).
245 Ibid.
246 Ibid.
247 Rosenberg, K. & Trevathan, W. (1995). Bipedalism and Human Birth: The Obstetrical Dilemma Revisited. *Evolutionary Anthropology*, 4 (5).
248 Mitteroecker, P. & Fischer, B. (2023). Evolution of the human birth canal. *American Journal of Obstetrics & Gynecology*
249 Garlinghouse, T. (2019). *Unraveling the Mystery of Human Bipedality*. Sapiens. https://www.sapiens.org/archaeology/human-bipedality/
250 Rosenberg, K.R. & Trevathan, W.R. (2001). The Evolution of Human Birth. *Sci. Am.*, 285(5).
251 Mitteroecker, P. & Fischer, B. (2023). Evolution of the human birth canal. *American Journal of Obstetrics & Gynecology*.
252 Rosenberg, K.R. & Trevathan, W.R. (2001). The Evolution of Human Birth. *Sci. Am.*, 285(5).
253 Rosenberg, K.R., Trewathan, W.R. (2021): The obstetrical dilemma revisited – revisited. In Han, S. & Tomori, C. (eds.),*The Routledge Handbook of Anthropology and Reproduction*. Routledge.
254 Lund, P.J.A. (2006). Semmelweis – en varsler. *Tidsskrift for Den norske legeforening*, 126(13–14). (In Norwegian. Alternative source in English: Zoltán, I. (08.04.24). *Ignaz Semmelweis. Encyclopedia Britannica*. https://www.britannica.com/biography/Ignaz-Semmelweis).
255 Skålevåg, S.A. (2020). *Ignaz Semmelweis* i Store norske leksikon. https://snl.no/Ignaz_Semmelweis (in Norwegian. Alternative source in English: Zoltán, I. (08.04.24). *Ignaz Semmelweis. Encyclopedia Britannica*. https://www.britannica.com/biography/Ignaz-Semmelweis).
256 Lie, S.O. (2000). *Merkesteiner i norsk medisin. Føllings sykdom*. Tidsskrift for Den norske legeforening. https://tidsskriftet.

no/2000/10/merkesteiner-i-norsk-medisin/follings-sykdom (in Norwegian. Alternative source in English: Britannica, T. Editors of Encyclopaedia (24.04.24). *phenylketonuria*. *Encyclopedia Britannica*. https://www.britannica.com/science/phenylketonuria).

257 Eberhard-Gran, M., Nordhagen, R., Heiberg, E., Bergsjø, P. & Eskild, A. (2003). Barselomsorg i et tverrkulturelt og historisk perspektiv. *Tidsskrift for Den norske legeforening*, 123 (24). (In Norwegian. The title translates as 'Neonatal care – a cross-cultural and historic perspective').

258 Schjødt, B. (2006, 1.01). *Norsk Barnesmerteforening stiftet*. Tidsskrift for norsk psykologisk forening. https://psykologtidsskriftet.no/nyheter/2006/01/norsk-barnesmerteforening-stiftet (in Norwegian. Alternative source in English: Rodriguez McRobbie, L (29.07.17) When babies felt no pain. *Boston Globe*. Retrieved 13.05.24 from: https://www.bostonglobe.com/ideas/2017/07/28/when-babies-felt-no-pain/Lhk2OKonfR4m3TaNjJWV7M/story.html)

259 John Bowlby popularised his attachment theory in the period 1970–1980.

260 Rosenberg, K.R., Trewathan, W.R. 2021: The obstetrical dilemma revisited – revisited. In Han, S. & Tomori, C. (eds.),*The Routledge Handbook of Anthropology and Reproduction*. Routledge.

261 Bohren, M.A., Hofmeyr, G.J., Sakala, C., Fukuzawa, R.K. & Cuthbert, A. (2017). Continuous support for women during childbirth. *Cochrane Database Syst.*, 7, CD003766. NB: some of the studies included in the review contain low-quality evidence.

262 Kongelstad, M. (2019). *Doula i Store medisinske leksikon*. https://sml.snl.no/doula (in Norwegian. Alternative source in English: Corfield, J. (01.05.24). *doula. Encyclopedia Britannica*. https://www.britannica.com/topic/doula)

263 Oslo Universitetssykehus (22.02.2021). Flerkulturell doula. https://oslo-universitetssykehus.no/likeverd-og-mangfold/

flerkulturell-doula (in Norwegian. Alternative source in English: Haugaard, A., Larsson Tvedte, S., Stene Severinsen, M., Henriksen, L. (2020) Norwegian multicultural doulas' experiences of supporting newly-arrived migrant women during pregnancy and childbirth: A qualitative study. *Sexual & Reproductive Healthcare.* 26:100540).

264 Lund, P.J.A. (2006). Semmelweis – en varsler. *Tidsskrift for Den norske legeforening*, 126(13–14). (In Norwegian. Alternative source in English: Zoltán, I. (08.04.24). *Ignaz Semmelweis. Encyclopedia Britannica.* https://www.britannica.com/biography/Ignaz-Semmelweis).

265 Thomassen, A.I. (12.04.2021). *Nok penger.* Barseloppøret. https://barselopproret.no/fordypning/nok-penger (in Norwegian. The title translates as "Enough money", and the document is a call to action to demand a rethinking of how maternity services in Norway are provided.)

266 Different sources give different pregnancy lengths for emperor scorpions, from seven months to over a year, and length of pregnancy probably varies with temperature, access to food, humidity and/or other environmental factors.

267 Ortega, R.P. (2020, 16.01). *This is the oldest scorpion known to science.* Science. https://www.science.org/content/article/oldest-scorpion-known-science

268 Scorpions can be divided into two groups, apoikogenic and katoikogenic, depending on how they provide nutrients to their foetuses. Apoikogenic scorpions have a kind of egg yolk, while katoikogenic scorpions do not – they provide their foetuses with nutrients via a placenta-like organ. See e.g. Volschenk et al. (2008) Comparative anatomy of the mesosomal organs of scorpions (Chelicerata, Scorpiones), with implications for the phylogeny of the order. *Zoological Journal of the Linnean Society*, 154.

269 Nesheim, B.-I. (2022). *Morkaken* i Store medisinske leksikon. https://sml.snl.no/morkaken (in Norwegian. Alternative source in English: Britannica, T. Editors of Encyclopaedia (02.05.24). *placenta. Encyclopedia Britannica*. https://www.britannica.com/science/placenta-human-and-animal).

270 Mota-Rojas, D., Orihuela, A., Strappini, A., Villanuea-García, D., Napolitano, F., Mora-Medina, P., Barrios-García, H.B., Herrera, Y., Lavalle, E. & Martínez-Burnes, J. (2020). Consumption of Maternal Placenta in Humans and Nonhuman Mammals: Beneficial and Adverse Effects. *Animals*. doi:10.3390/ani10122398.

271 Nesheim, B.-I. (2022). *Morkaken* i Store medisinske leksikon. https://sml.snl.no/morkaken (in Norwegian. Alternative source in English: Huffman, J. W. (11.05.24). *pregnancy. Encyclopedia Britannica*. https://www.britannica.com/science/pregnancy).

272 Mota-Rojas, D., Orihuela, A., Strappini, A., Villanuea-García, D., Napolitano, F., Mora-Medina, P., Barrios-García, H.B., Herrera, Y., Lavalle, E. & Martínez-Burnes, J. (2020). Consumption of Maternal Placenta in Humans and Nonhuman Mammals: Beneficial and Adverse Effects. *Animals*. doi:10.3390/ani10122398.

273 Oksman, O. (10.02.2016). *Eating your placenta – is it healthy or just weird?* The Guardian. https://www.theguardian.com/lifeandstyle/2016/feb/10/eating-your-placenta-healthy-motherhood-new-mothers-infants-postpartum-depression-placentophagy-fda

274 Renfree, M.B. (2010). Review: Marsupials: Placental Mammals with a Difference. *Placenta*, 24.

275 Ostrovsky, A.N., Lidgard, S., Gordon, D.P., Schwaha, T., Genikhovich, G. & Ereskovsky, A.V. (2016). Matrotrophy and placentation in invertebrates: a new paradigm. *Biol. Rev. Camb. Philos. Soc.*, 91(3).

276 Renfree, M.B. (2010). Review: Marsupials: Placental Mammals with a Difference. *Placenta*, 24.

277 Sadedin, S. (2014, 4.08). *War in the womb*. Aeon. https://aeon.co/

essays/why-pregnancy-is-a-biological-war-between-mother-and-baby

278 Callier, V. (2015, 2.09.). *Baby's Cells Can Manipulate Mom's Body for Decades*. Smithsonian Magazine. https://www.smithsonianmag.com/science-nature/babyscells-can-manipulate-moms-body-decades-180956493/

279 Mayer, G., Franke, F.A., Treffkorn, S., Gross, V. & Oliveira, I. de S. (2015). Onychophora in Wanninger, A. (ed.). *Evolutionary Developmental Biology of Invertebrates 3: Ecdysozoa I: Non-Tetraconata*. Springer-Verlag.

280 Wright, J. (2014). *Onychophora*. Animal Diversity Web. https://animaldiversity.org/accounts/Onychophora/

281 Tait, N.N. & Norman, J.M. (2001). Novel mating behaviour in Florelliceps stutchburyae gen. nov., sp. nov. (Onychophora: Peripatopsidae) from Australia. *J. Zool. Lond.*, 253.

282 Ibid.

283 The sand tiger shark is pregnant for between nine and twelve months. Bansemer, C.S. & Bennett, M.B. (2009). Reproductive periodicity, localised movements and behavioural segregation of pregnant Carcharias taurus at Wolf Rock, southeast Queensland, Australia. *Marine Ecology Progress*, 374.

284 Wikipedia (n.d.). *Sand tiger shark*. Retrieved 28.11.22 from https://en.wikipedia.org/wiki/Sand_tiger_shark

285 This phenomenon of embryos cannibalising other embryos is called embryophagy. The behaviour of the African social spider we met previously, where the young eat their mother, is called matriphagy.

286 Gilmore, R.G., Putz, O. & Dodrill, J.W. (2005). Oophagy, Intrauterine Cannibalism and Reproductive Strategy in Lamnoid Sharks. In Hamlett, W.C. (ed.) *Reproductive Biology and Phylogeny of Chondrichtyes*. Science Publishers Inc.

287 Greven, Helmut, Professor Emeritus at Heinrich Heine University, Düsseldorf. Personal correspondence, 10.06.22.

288 Smithsonian Institution (2018). *Coelacanth*. Ocean. https://ocean.si.edu/ocean-life/fish/coelacanth

289 NOAA Fisheries (2022). *Sperm Whale*. Species directory. https://www.fisheries.noaa.gov/species/sperm-whale

290 Havforskningsinstituttet (2021). *Tema: Hummer – europeisk*. Havforskningsinstituttet. https://www.hi.no/hi/temasider/arter/hummer-europeisk (in Norwegian. Alternative source in English: Britannica, T. Editors of Encyclopaedia (27.04.24). lobster. *Encyclopedia Britannica*. https://www.britannica.com/animal/lobster).

291 Montague, M. (2021). *Elephant gestation period longer than any living mammal*. BBC Earth. https://www.bbcearth.com/news/elephant-gestation-period-longer-than-any-living-mammal

292 López-Romero, F.A., Klimpfinger, C., Tanaka, S. & Kriwet, J. (2020). Growth trajectories of prenatal embryos of the deep-sea shark Chlamydoselachus anguineus (Chondrichthyes). *J. Fish Biol.*, 97.

293 Long, J.A., Trinajstic, K., Young, G.C. & Senden, T. (2008). Live birth in the Devonian period. *Nature*, 453.

294 Voje, K.L. (2022). *Livets evolusjonshistorie* i Store norske leksikon. https://snl.no/livets_evolusjonshistorie (in Norwegian. Alternative source in English: Ayala, F. Jose (23.04.24). evolution. *Encyclopedia Britannica*. https://www.britannica.com/science/evolution-scientific-theory).

295 Blackburn, D.G. (2006). Squamate reptiles as model organisms for the evolution of viviparity. *Herpetological Monographs*, 20.

296 Delsett, L.L. (2021). *Fiskeøgler* i Store norske leksikon. https://snl.no/fiskeøgler (Britannica, T. Editors of Encyclopaedia (11.05.24). ichthyosaur. *Encyclopedia Britannica*. https://www.britannica.com/animal/ichthyosaur).

297 Blackburn, D.G. (2015). Evolution of Vertebrate Viviparity and Specializations for Fetal Nutrition: A Quantitative and Qualitative Analysis. *Journal of Morphology*, 276.

298 Around now in evolutionary terms means April 2019 in this case.
299 Voje, K.L. (2022). *Livets evolusjonshistorie* i Store norske leksikon. https://snl.no/livets_evolusjonshistorie. Where no reference is provided in this timeline, the information is taken from this source.
300 Emera, D., Romero, R. & Wagner, G. (2012). The evolution of menstruation: A new model for genetic assimilation. *BioEssays*, 34(1).
301 Frontiers in Ecology and Evolution (no date). *Origin and Early Evolution of Amniotes*. Research Topic. Retrieved 15.09.22 from https://www.frontiersin.org/research-topics/14947/origin-and-early-evolution-of-amniotes
302 Otto, S. (2008). Sexual Reproduction and the Evolution of Sex. *Nature Education* 1(1).
303 Wiktionary (n.d.). *Embryophagy*. Retrieved 22.04.24 from https://en.wiktionary.org/wiki/embryophagy
304 Fenelon, J.C. & Banerjee, A. (2014). Embryonic diapause: development on hold. *Int. J. Dev. Biol.*, 58.
305 Wikipedia (n.d.). *Hemipenis*. Retrieved 10.12.22 from https://en.wikipedia.org/wiki/Hemipenis
306 Wikipedia (n.d.). *Hermaphrodite*. Retrieved 10.12.22 from https://en.wikipedia.org/wiki/Hermaphrodite
307 Young, T.P. (2010). Semelparity and Iteroparity. *Nature Education Knowledge*, 3(10).
308 Wikipedia (n.d.). *Matriphagy*. Retrieved 10.12.22 from https://en.wikipedia.org/wiki/Matriphagy
309 Young, T.P. (2010). Semelparity and Iteroparity. *Nature Education Knowledge*, 3(10).
310 Wikipedia (n.d.). *Squamata*. Retrieved 25.04.24 from https://en.wikipedia.org/wiki/Squamata
311 Kiserud, T. (2012). How long does a pregnancy last? *Tidsskrift for Den norske legeforening*, 132.

Index of Species

African bush elephant (*Loxodonta africana*) 173–174, 180, 234, 241
African social spider (*Stegodyphus dumicola*) 13, 66–67, 181, 239, 251
Antechinus 130
Atlantic wolffish (*Anarhichas lupus*) 163–167, 238
Australian brush-turkey (*Alectura lathami*) 38–39, 156, 166

Beaver (*Castor fiber*) 122–123, 181, 183
Bed bug (*Cimex lectularius*) 32–33
Blue-footed booby (*Sula nebouxii*) 172
Bonobo (*Pan paniscus*) 31
Brown bear (*Ursus arctos*) 51, 59

Brown rat (*Rattus norvegicus*) 110
Brown-throated sloth (*Bradypus variegatus*) 159–162

Chimpanzee (*Pan troglodytes*) 65, 192, 243
Cryptasterina hystera 30
Clonal Plumose Anemone (*Metridium senile*) 15–16, 18, 24, 71, 96, 239
Coelacanths (Coelacanthiformes)
Common bottlenose dolphin (*Tursiops truncatus*) 203–204, 233
Common eider (*Somateria mollissima*) 48–51, 71, 96, 238
Common pill woodlouse (*Armadillidium vulgare*) 53–54

295

Common shrew (*Sorex araneus*) 121
Cordyceps 68–69
Ctenochromis horei 19–21
Cuckoo catfish (*Synodontis multipunctatus*) 20–21, 158

Damaraland mole-rat (*Cryptomys damarensis*) 110–114, 239
Dolphin (Delphinidae) 31, 224

Eastern grey kangaroo (*Macropus giganteus*) 58–60, 71, 238, 243
Emperor penguin (*Aptenodytes forsteri*) 25–27, 78–80, 165, 181
Emperor scorpion (*Pandinus imperator*) 12, 217–220, 238
Escherichia coli 17, 98, 105
European fire salamander (*Salamandra salamandra bernardezi*) 233, 235–236
European lobster (*Homarus gammarus*) 234

Frilled shark (*Chlamydoselachus anguineus*) 234, 235

Geoffroy's tailless bat (*Anoura geoffroyi*) 130–133, 193
Giant Pacific octopus (*Enteroctopus dofleini*) 13, 73–75, 147–150, 239, 253
Gorilla (*Gorilla gorilla*) 206–207
Green acouchi (*Myoprocta pratti*) 116–119
Guinea pig (*Cavia porcellus*) 76–78

Hanuman langur (*Semnopithecus* spp.) 168–171, 241
Hippopotamus (*Hippopotamus amphibius*) 195–198
Hoverfly (Syrphidae) 32
Hypacrosaurus stebingeri 156–157

Ichthyosaur (Ichthyosauria) 236–237

Jonah's icefish (*Neopagetopsis ionah*) 154–155

Komodo dragon (*Varanus komodoensis*) 36–38, 97, 174, 199–201

Laysan albatross (*Phoebastria immutabilis*) 93, 173
Lion (*Panthera leo*) 169
Little spotted kiwi (*Apteryx owenii*) 81–86, 238, 256

Lophelia pertusa 18–19, 238, 248

Maiasaura peeblesorum 157–158
Moth (Microlepidoptera) 161–162

Namaqua chameleon (*Chamaeleo namaquensis*) 56–58, 142–143, 239
Nile crocodile (*Crocodylus niloticus*) 105–109, 158, 183

Ocellaris clownfish (*Amphiprion ocellaris*) 98–99
Orca (*Orcinus orca*) 80, 177–181

Pacific beetle cockroach (*Diploptera punctata*) 86, 238
Peacock (*Pavo* spp.) 94, 100
Placoderms 236
Platypus (*Ornithorhynchus anatinus*) 52–53, 72, 246

Rabbit (*Oryctolagus cuniculus*) 55–56, 183
Reindeer (*Rangifer tarandus*) 186–189
Reindeer botfly (*Hypoderma tarandi*) 188–189

Salmon (*Salmo* spp.) 149, 177–178

Sand tiger shark (*Carcharias taurus*) 231–232, 235–236, 238
San Martin titi (*Plecturocebus oenanthe*) 134–136, 181, 243
Sheep (*Ovis aries*) 144–146
Sheep bot fly (*Oestrus ovis*) 145–146
Southern Darwin's frog (*Rhinoderma darwinii*) 87, 140
Sperm whale (*Physeter macrocephalus*) 234
Split-gill mushroom (*Schizophyllum commune*) 89
Spotted hyena (*Crocuta crocuta*) 12, 124–127, 239
Springtail (Collembola) 32
Surinam toad (*Pipa pipa*) 137–140

Thomson's gazelle (*Eudorcas thomsonii*) 152–153
Triplewart seadevil (*Cryptopsaras couesii*) 35

Quadrartus yoshinomiyai 173, 174–175, 239

Velvet worm (Onychophora) 226–229

Virginia opossum (*Didelphis virginiana*) 29–30, 31–32, 192

Wandering albatross (*Diomedea exulans*) 88, 104

Water flea (Cladocera) 97–98, 174

White's seahorse (*Hippocampus whitei*) 40–42, 140, 166, 181

Wolf (*Canis lupus*) 111

ANNA BLIX was inspired to write this book by her experience during her first pregnancy: "I was incredibly sick. I vomited throughout my entire pregnancy and had to lie in a dark room for the entire first trimester just to be able to keep anything down. I couldn't look at screens, couldn't read books, and could barely even listen to the radio. As a biologist, I couldn't understand how this could be possible." Having previously worked at the Norwegian parliament, Anna is now a senior advisor at WWF Norway. She writes a regular environmental column for the newspaper *Klassekampen*.

NICHOLA SMALLEY's translation of Andrzej Tichý's novel *Wretchedness* won the 2021 Oxford-Weidenfeld Prize and was shortlisted for the Bernard Shaw Prize that same year. It was also longlisted for the 2021 International Booker Prize. Her translation of *A System So Magnificent It Is Blinding* by Amanda Svensson was longlisted for the International Booker Prize 2023.